Homo Problematis Solvendis—Problem-solving Man

From the Original Drawing by Leonardo da Vinci.   Engraved by F. Bartolozzi R.A. Historical Engraver to his Majesty.

IN HIS MAJESTY'S COLLECTION.

Publish'd as the Act directs May 13 1796 by I. Chamberlaine

David H. Cropley

Homo Problematis Solvendis —Problem-solving Man

A History of Human Creativity

 Springer

David H. Cropley
School of Engineering
University of South Australia
Adelaide, SA, Australia

ISBN 978-981-13-3100-8   ISBN 978-981-13-3101-5   (eBook)
https://doi.org/10.1007/978-981-13-3101-5

Library of Congress Control Number: 2018961724

Cover image: 'Engineering: a hoist in use at an arsenal. Engraving by F. Bartolozzi, 1796, after Leonardo da Vinci.' by Leonardo da Vinci. *Credit* Wellcome Collection. CC BY https://wellcomecollection.org/works

This Springer imprint is published by the registered company Springer Nature Singapore Pte Ltd.
The registered company address is: 152 Beach Road, #21-01/04 Gateway East, Singapore 189721, Singapore

Preface

This book is the culmination of a number of years of scholarship and research in the fields of engineering and creativity. When I tell people that I do research in creativity and explain that I work in a university engineering school, I frequently get some puzzled looks. "What has creativity got to do with engineering?" I sometimes groan inwardly but then explain that creativity is all about generating new and effective ideas, while the essence of engineering is finding solutions to problems, and people generally start to get it. In fact, most people, at this point, start to get quite animated and want to give examples—"Oh yeah, that's like Edison and the electric light!" or similar. This always reassures me and tells me that people generally have a pretty good, if unconscious, understanding of what engineering is all about. Not only that, but most people understand just how widespread and important engineering is, and has been, in their own lives. Engineers solve problems—we call it *design*, but its essence is harnessing the materials and forces of nature for defined purposes. Creativity is all about solving problems, and the two fields are therefore intimately intertwined. As I hope to show in this book, humans are not only remarkably good engineers, they are remarkably *creative* engineers. If that wasn't the case, I'd probably currently be sitting in a dark cave grunting at you and gnawing on a bone. To me, engineering is the ultimate expression of creativity—the process of finding new and effective solutions to problems ranging from finding food, shelter, security and safety all the way up to landing humans on the Moon.

Of course, not all of the inventions I present in this book are the sole domain of engineers. There are also some medical inventions, literary inventions and even economic or social inventions. However, all can be described as broadly *technological*, in the sense that they involve the application of science, or scientific knowledge, and also benefit society in some way. An alternative title for this book could therefore be "A History of Human *Technological* Creativity" and that would be broad enough to cover any process of applying specialist knowledge for some beneficial, problem-solving purpose.

Another important goal of this book is also to get the reader to think more deeply about what we mean by creativity. The term is frequently overused and misapplied with the consequence that there are many myths and misunderstandings surrounding creativity. Creativity is only found in artistic pursuits? Creativity can't be taught—you are either born with it, or you are out of luck! Creativity can't be defined. Creativity is a mysterious gift from the Gods! I hope that, through the process of examining the inventions in this book, the reader will see not only that creativity *can* be found everywhere but that it is entirely understandable, and able to be developed, manipulated and improved.

Creativity is, in fact, four interrelated things. It is an attitude, or a mindset, of individuals. Some people are motivated to seek out new and useful things and have personal dispositions that favour new experiences. Creativity is also a way of thinking. The ability to generate many *possible* solutions to a problem—divergent thinking—is frequently seen as the defining essence of creativity. Creativity is also impacted by the *environment* in which these activities take place. Individuals may be highly motivated and good divergent thinkers, but if the system acts against them, and is unsupportive, it may still be difficult to generate creative outcomes. Finally, creativity is also a property of the things created. For a product—an invention—to be creative, it must be, at the very least, new and effective.

Through the course of ten *ages*, and across 30 different inventions, I hope to show you all of these facets of creativity in action. However, a particular focus for us will be the inventions themselves—why should these products, devices and systems be regarded as *creative*?

Adelaide, Australia   David H. Cropley

Acknowledgements

This book is dedicated to my mother, Alison, who passed away unexpectedly just before I started developing the concept. She could cook, sew, knit, crochet, paint, make lace, make pottery, teach and do a thousand other practical, skilful things. I suspect that without her, and millions of creative, problem-solving women and mothers before her, we'd all still be living in caves, hungry, cold and angry. Thank you!

I would also like to thank my father, Arthur—creativity researcher *par excellence*—for his invaluable technical insights and knowledge of creativity. My wife Melissa, and children, Matthew, Dana and Daniel, are pillars of creative expression in writing, acting, theatre and film, and they always support me.

One of humankind's greatest achievements is no doubt the ingenuity and creativity that we have applied to the problems of medicine and health care. While writing this book, I have twice benefitted from this after suffering from serious eye problems. I'd like to thank all of the doctors, nurses, biomedical engineers, IT specialists and other people who are part of the new Royal Adelaide Hospital in South Australia. I feel very lucky to live in a country where world-class, universal health care is prioritised.

Contents

List of Figures

List of Tables

Introduction

Nobody really knows what creativity is! Some say it's a gift from the Gods; others ascribe their creativity to their "*Muse*".[1] Many insist that it can't be defined, and certainly not sullied by something as crass as measurement! It is perhaps no surprise, then, that many people associate creativity exclusively with artistic pursuits. Even the word—creativity—is often used as shorthand for the arts, with many in these professions labelling themselves as *creatives*. Of course, there is nothing wrong with this sort of linguistic appropriation, except that it reinforces certain myths and stereotypes, and leaves little room for manifestations of creativity in other, non-artistic domains.

So if you picked up this book expecting to read about the history of dance, poetry and literature, you are going to be disappointed. This is not to say that those pursuits are not creative, or not worthy of discussion. Rather, it is to draw attention to the common, underpinning nature of creativity in any activity—namely, *the production of effective novelty*[2]—and to show how modern humans have used their remarkable capacity for finding new solutions to new problems—creative problem solving—to thrive on a planet that is in a constant state of change.

The subject of this book is a presentation of the history of the creativity and innovation of modern humans, through examples of solutions to the range of human needs that have been developed over many centuries. The title—*Homo problematis solvendis*—is a play on the system of scientific classification of human species

---

[1] The Muses were the Goddesses of ancient Greece whose attentions supposedly inspired the production of poetry, literature, art and science. Each had a speciality—thus Terpsichore stimulated dance, Polyhymnia kindled the production of sacred hymns, and Urania encouraged astronomers. In modern times, we often describe someone who moves us to artistic and literary endeavours as our *Muse*.

[2] See, for example, Cropley, D. H. (2015). *Creativity in engineering: Novel solutions to complex problems*, San Diego, CA: Academic Press.

© Springer Nature Singapore Pte Ltd. 2019
D. H. Cropley, *Homo Problematis Solvendis—Problem-solving Man*,
https://doi.org/10.1007/978-981-13-3101-5_1

(e.g. *Homo habilus*, *Homo erectus*) and is intended to suggest that a defining characteristic of *modern* humans is not so much our *wisdom* (i.e. *Homo sapiens*) as it is our fundamental ability to solve problems (hence *Problem-solving man* = Homo problematis solvendis). You will not find my classification in any scientific texts on taxonomy, and it is intended primarily as a literary device, and to help us examine ourselves in a different and unconventional light. I should also add that, while I've used the scientific, Latin term *Homo*, the focus of this book is very much a celebration of the creativity of millennia of pioneering women and men, whose ingenuity has overcome all manner of challenges—many literally matters of life and death—and without whom, none of us might exist today.

The book will address the history of human creativity and innovation first by explaining what creativity and innovation are, and why human "needs" act as the stimuli to problem solving (i.e. creativity/innovation). The book will then explore innovations over ten distinct "ages" of human history, beginning with "Prehistory", and moving up to the present "Digital Age". Each era of human history will be covered by one chapter, with three key innovations of that era described in each chapter. Unlike other books that discuss and describe inventions and human ingenuity,[3] this book focuses not merely on "what" was invented, or "who" did the inventing, but on "why" it was invented, and "why" it should be considered *creative*. What need did each innovation satisfy, and how have humans drawn on their innate problem-solving ability—their capacity for creativity/innovation—to satisfy these needs? In this manner, the book is a history of the psychological capacity of humans to identify and solve problems (creativity and innovation), and not simply a catalogue of the history of technology, or a biography of inventors.

The innovations selected for each chapter have been chosen because they represent *designed* solutions, i.e. deliberately invented solutions to problems, as opposed to mere accidental discoveries. For this reason, I do not include *fire* as one of the innovations. Although it is impossible for us to know with any real certainty, it seems highly likely, and scholars suggest, that fire was first controlled, probably by one of our evolutionary ancestors, Homo erectus, as much as 1.7 million years ago. *Controlled by* is also important—fire itself would have occurred naturally, due to phenomena such as lightning strikes. However, the ability to deliberately create fire, for example using flints, or fire sticks, while a purposeful invention, predates our focus on modern humans—it appears that we have another species to thank for that!

In addition to the criterion of *designed* solutions, the innovations have been chosen to favour those that satisfy basic human needs—fundamental problems—such as the need for food, shelter, safety, rest, security, transport and so forth. Sadly, this means that fidget spinners and yoyos are out (even if you are tempted to claim that they satisfy a psychological need for a feeling of accomplishment)! We can draw on Abraham Maslow's so-called *hierarchy of needs*[4] as another piece in the puzzle of human creativity and problem solving. Maslow theorised that people are motivated

---

[3]I recommend, for example, Melissa Schilling's book *Quirky* (2018) published by Public Affairs (New York), or Amina Khan's *Adapt* (2017) published by Atlantic Books (London).

[4]Maslow, A. H. (1943). A theory of human motivation. *Psychological Review*, *50*(4), 370–396.

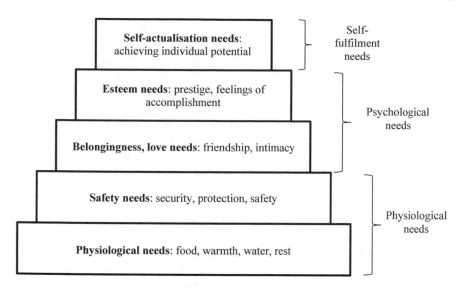

Fig. 1 Maslow's hierarchy of needs (based on Maslow, A. H. (1943). A theory of human motivation. *Psychological Review, 50*(4), 370–396)

to satisfy certain needs—what I'm generally referring to as source problems—and that these needs/problems have a certain hierarchy, or precedence, to them. Thus, the most basic *problems* that we must solve are those at the base of the pyramid (see Fig. 1)—how to satisfy our hunger, our need to breathe, our need to keep warm/cool and so on. Once those basic needs are met—that is, the basic problems are solved—as humans we then start to turn our minds to needs/problems higher up the pyramid. If I differ from Maslow, it is only to observe that even the solved problems, or the satisfied needs, rarely stay solved for long. Mother Nature rarely allows us the luxury of getting too comfortable. Maslow's theory therefore explains one key driver for human creativity, with *change* explaining another. Our inherent desire to move up the hierarchy, coupled with externally imposed changes, have kept our problem-solving creativity sharply honed for millennia.

In fact, there is a closer connection between creativity and Maslow's Hierarchy of Needs than may be apparent from more recent creativity research. Maslow (1974),[5] for example, discusses creativity in the context of what he calls *self-actualising people*, and distinguishes these individuals from the creativity of *special-talent creativeness* (i.e. eminent individuals such as painters, poets and artists).

It seems that Maslow, to some extent at least, regarded self-actualisation—the top of his hierarchy—as more or less as synonymous with creativity. Creativity, in other

---

[5]Maslow, A. H. (1974). Creativity in self-actualizing people. *Readings in human development: A humanistic approach*, 107–117. In fact, the source of this reference is a lecture Maslow gave in February, 1958—itself interesting in the history of the modern era of creativity. This lecture took place only months after the launch of Sputnik I—the world's first artificial satellite—that event being seen as the spark of the modern interest in creativity.

words, is self-actualisation. While Maslow himself understood that creativity was *not* synonymous with art (a persistent misconception, even today[6])—creativity can be found outside of the production of artistic works—it seems that he switched his attention away from products altogether and did not explore the idea that creativity could be found in other kinds of outcomes or artefacts. Tying these ideas together, we should understand that creativity is expressed at *every level* in Maslow's hierarchy. At the highest level, creativity is how we solve the problem of self-actualisation, and we can call this *transcendental creativity*. In the middle of the hierarchy, creativity is how we solve the problems of esteem and a desire to belong. We can call this *social creativity*. Finally, at the lowest level of the hierarchy, creativity is how we solve the problems of a need for safety, food, shelter and so forth. The creativity embodied in solving these basic needs is what we can call *functional creativity*.[7]

Creativity therefore is both the *means* and the *end*—it is a tool for solving problems and an end state in itself. In the same way that you don't seek intelligence as a goal in itself—you seek to develop intelligence because it adds value to your life by helping you to do things—so we seek creativity, not merely to label ourselves as creative but because creativity is a vital component of solving challenging problems and responding to needs.

In fact, this end state—creativity—can be found both in people and in things, and our focus in this book is very much on the latter. This also opens up another question. What are the things that can be regarded as creative? In fact, the inventions in this book are not confined to tangible artefacts—i.e. things that you can hold, touch and feel. They include *processes* (methods for achieving tangible or intangible outcomes—e.g. a production line in a factory), *systems* (complex interactions of hardware, software and people—e.g. the World Wide Web) and *services* (organised, but usually intangible, systems of labour and material aids designed to meet a need—e.g. Education). We will seek examples of the diverse nature of creative outcomes as we explore our catalogue of inventions.

Why Do We Need Creativity?

Since J. P. Guilford[8] first called for more attention on creativity as a way that human intellectual performance is expressed, there have been a number of waves of interest in

---

[6]We only need to look at the cover image of a Special Issue of TIME Magazine, published in August 2018, to see that this myth is alive and well. The cover tells us that it is a special issue on the science of creativity but shows an image of the two hemispheres of the human brain, with the right hemisphere brightly splashed with many colours, while the left hemisphere is blank. Not only does this reinforce the myth that creativity is art but it also suggests that creativity is confined to the right hemisphere—a notion also debunked in recent years by brain imaging studies.

[7]See for example, Cropley, D. H. and Cropley, A. J. (2005). Engineering creativity: A systems concept of functional creativity. In J. C. Kaufman and J. Baer (Eds.), *Creativity Across Domains: Faces of the Muse*, Chapter 10 (pp. 169–185). Hillsdale, NJ: Lawrence Erlbaum Associates Inc.

[8]Guilford, J. P. (1950). Creativity. *American Psychologist, 5*, 444–454.

creativity. At least several of these have been strongly tied to education. Guilford's initial call started a burst of research exploring creativity in an educational and developmental context. A wave of activity then followed the launch of Sputnik I by the Soviet Union in 1957.[9] That surge of interest, sparked by the perceived threat of Soviet superiority in the Space Race, died down once the United States effectively won that competition after landing the first humans on the moon in July 1969. There have been other, regular, bursts of enthusiasm for creativity in education over the last 50 years, however, it seems that none has ever really stuck, perhaps until now.

My theory explaining this lies in the source of the interest in creativity. Any process of innovation—developing new and effective solutions to a defined problem—requires two things. It requires a problem, or need, and it requires a solution. When the driver in this process comes from an expressed need—what is often referred to as *market pull*—we see solutions developed targeting this need and, more often than not, the problem is solved. However, when the driver in this process begins with a solution—what we call *technology push*—we frequently see this as a situation in which we have a solution for which there is no need. The difficulty in this situation is that, unless we can find a need, we have something that nobody wants.

I think this is why some waves of creativity have failed to stick over the last 50 years. Sometimes creativity experts come to the market selling creativity—technology push, in other words—but don't really articulate why the customer needs it. In this situation, we have a solution, but no real problem. Not surprisingly, after a while, interest wanes because the loop is not closed. Problems need solutions, just as solutions need problems. It's not enough simply to say that creativity is good for us, or even that it is important in education. Human nature demands a more compelling reason.

The times in the past when interest in creativity has persisted have been because the problem was more clearly articulated. Rather than creativity researchers and experts saying "you need creativity", the market has initiated the conversation by saying "we need creativity!" I believe that we are entering such a period again, and it is being driven by technology. Industry 4.0 is a term that you will hear more and more in the future. It describes the rise of so-called cyber-physical systems—combinations of artificial intelligence, digitisation, smart systems, the Cloud, big data and so forth—that are now being realised and exploited in industry. This is driving interest in creativity as a work skill because, in a world of Industry 4.0 and cyber-physical systems, the jobs of the future will be those things that computers *can't do*. For the time being, at least, that remains the very human ability to generate new and effective solutions to problems arising from human needs—creativity!

So why do we need creativity—the ability to generate new and effective ideas—and why do we need innovation—the ability to make use of those creative ideas? My *go-to* explanation is as follows. Change is the key. By that, I mean things like climate change, economic change, demographic change, technological change and so forth. Throughout human history, we have been subject to these sorts of

[9]See Cropley, D. H. (2015). *Creativity in engineering: Novel solutions to complex problems*, San Diego, CA: Academic Press, for a discussion of creativity, engineering and the Sputnik Shock.

changes, and every change that occurs has one of two basic effects. On the one hand, change—for example, a rise in atmospheric carbon leading to an increase in extreme weather events—generates new *demands* from people. These demands are sometimes referred to as *Market Pull*. We demand, for instance, new, carbon-free forms of electricity generation, or we demand cars powered by batteries and not internal combustion engines. These new demands define new problems that must be solved. On the other hand, change gives rise to new solutions—e.g. vortex reducing *winglets* on the wing tips of commercial aircraft that reduce drag, and therefore reduce fuel consumption. Whichever path arises from change, innovation (underpinned by creativity) is the process of connecting a new problem to a new solution.

Creativity is needed because it is the key to developing and dealing with newness or novelty. Change generates new problems and gives rise to new solutions. The newness inherent in each requires the capacity for creativity, while connecting the problems and solutions together is the process of innovation. As long as we are dealing with novelty in the context of problem solving, we cannot escape the need for creativity. It is essential therefore, especially in an era of Industry 4.0, that we understand what creativity is, why it drives innovation, and also how we can maximise this vital, twenty-first-century skill.

Ten *Ages* of Innovation

The different "Ages" used in this book are intended to be broadly representative of different eras of human development. They are not intended to be a full and complete timeline, nor should they exclude other periods. There is also some fuzziness and overlap, as well as some gaps, meaning that they should be taken more as *signposts* as opposed to strict labels. They are intended simply to give the reader a sense of the character of the period in question, as well as its place in the broader span of human development. Although the ages cover different spans of time, they do give a sense of the accelerating pace of change and innovation as we move forward in time. One note about dates—I will use the designation "BCE" to mean *Before the Common Era*, and "CE" to refer to the *Common Era*. These have the same basic meaning as BC and AD but are religiously neutral. Normal practice is to specify when a date is BCE but to drop the designation for dates that are in the Common Era. I will use the suffixes for our early time periods, and once we are comfortably into the Common Era, I will drop their use.

The ten ages that we will focus on are:

(a) Prehistory, covering the period from the dawn of modern humans up to the invention of writing, around 3000 BCE;
(b) The Classical Period, spanning the era from approximately 800 BCE to 500 CE;
(c) The Dark Ages, taking us from about 500 CE up to 1450 CE;
(d) The Renaissance, spanning the period 1300 CE–1700 CE;
(e) The Age of Exploration, covering the years from about 1490 CE to 1799 CE;

(f) The Age of Enlightenment, which takes us from 1685 CE up to 1815 CE;

(g) The Romantic Period, roughly the period from 1800 CE to 1900 CE;

(h) The Modern Age, which was essentially the first half of the twentieth century, from 1900 CE to into the 1950s;

(i) The Space Age, covering a period from the late 1950s into the 1980s;

(j) The Digital Age, which we can think of as the current era, beginning in the 1980s.

Our focus in this book is on *modern humans*. In scientific terms, we (by which I mean every single one of us alive today) belong to the broader genus *Homo* ("human being" in Latin), and the more particular species *Homo sapiens*. By comparison, the four-legged creature that terrorised Little Red Riding Hood is a member of the genus *Canis* (i.e. dog) and the particular species *Canis lupus* (i.e. wolf). Within our parent genus,[10] there have been various species, some of which I have already mentioned. These include the extinct *Homo habilus* (handy/skilful man), and *Homo erectus* (upright man), and also our most recent relatives, *Homo neanderthalensis*. Modern humans, in the sense of Homo sapiens, are thought to have emerged[11] from the evolutionary process as early as 300,000 years ago, probably coexisting with other species (e.g. Neanderthals) until as late as 30,000 years ago. So-called *anatomically modern humans* (AMH)—us, in other words—have existed more or less as we are now, in both physiological and intellectual terms, since at least 30,000 years ago. The science of this topic continues, and new evidence is constantly emerging that updates our knowledge of the history and evolution of modern humans. However, it seems safe to say that the periods I consider in this book and the inventions in question are, without a doubt, the work of our direct, modern human ancestors.

Another important consideration is that the inventions I have selected did not necessarily make their *first* (or last) appearance in the epoch that I have placed them in. For example, there is ample evidence that prehistoric cultures made use of rudimentary calendars—think of Stonehenge for example—however, I introduce the calendar in the Classical Period (approximately the eighth century BCE to the fifth century CE) because we can identify what appears to be the first emergence of more sophisticated calendars based on the solar year in that time period. We should not get too bogged down with the strict definitions of these time periods—my purpose is to capture the essence of different epochs, and not catalogue dates and times. I'm an engineer, not a historian!

I ask the reader to remember that our primary focus is to explore the driving need behind the invention, and not necessarily a detailed history of the invention itself. For this reason, it is sometimes necessary to be selective in the exact detail of when an invention emerged. Where necessary, I will give some contextualising information and refer the reader to other sources for a more complete discussion of the invention itself.

---

[10]Swedish botanist and zoologist Carl Linnaeus (1707–1778) popularised this system of taxonomy, but French botanist Joseph Pitton de Tournefort (1656–1708) is regarded as its inventor.

[11]Another cautionary note here—if you don't believe in evolution, you might want to skip ahead.

In fact, this question of the exact *when* of an invention is consistent with the true nature of *invention*. In other words, there is invention both in an incremental sense—improving on what already exists—and there is invention in a radical sense—the emergence of never-before-seen inventions. In some cases, our inventions will be incremental in nature, and in other cases, radical.

For each innovation, I will follow the same basic format. *What was invented*—in other words, a basic summary of the innovation in question; *Why was it invented*—in particular, what need did it satisfy, or what problem did it solve, and; *How creative was it*—a more objective, systematic scoring of the creativity embodied in the invention. In each case study, I try to summarise the underpinning nature of the innovation by expressing its core purpose or function in the form of "How to verb noun?" In this latter case, I mean, for example, that the basic function of a screwdriver can be expressed as *How to apply torque.*[12] Describing any invention in this way is a good mechanism for understanding not only what it does but also what other ways might be developed to achieve the same function.

---

[12]Torque, very simply, is rotational force.

Measuring Creativity

A key feature of this book will be the analysis of the creativity of each of the 30 inventions chosen for analysis. This means that our focus is very much on the creativity of the product itself, as distinct from the personal qualities, thinking processes or environmental factors that also impact on creativity. An important part of the discussion of each invention will be the task of assigning it a creativity score—answering the question *how creative is it*? This will help us to understand and appreciate what problem the invention was trying to solve, and the driving force behind the invention.

Although some variation exists in the body of creativity research, there is a reassuringly high level of agreement about what makes something—*a product*—creative. Most researchers now agree that the creativity of an idea, solution or product must, as a minimum, be defined by:

- Relevance and effectiveness—does the product[1] do what it is supposed to do?
- Novelty—is the product new, original and surprising?

To explore this topic further, I recommend not only two of my own studies[2] but also the Creative Product Semantic Scale (CPSS) by Besemer and O'Quin.[3] My own research built on these to define additional criteria of:

- Elegance—is the product complete, fully worked out, understandable?
- Genesis—does the product open up new perspectives?

---

[1] I will generally use the word "product" or "artefact" in this book to mean the result of any creative, problem-solving process. This does not mean that the result can only be a tangible thing.

[2] Cropley, D. H. and Kaufman, J. C. (2012). Measuring Functional Creativity: Non-Expert Raters and the Creative Solution Diagnosis Scale (CSDS), *Journal of Creative Behavior*. 46:2, pp. 119–137. Cropley, D. H., Kaufman, J. C. and Cropley, A. J. (2011). Measuring Creativity for Innovation Management, *J. Technol. Manag. Innov. Vol. 6, No. 3*, pp. 13–30.

[3] Besemer, S. P., & O'Quin, K. (1987). Creative product analysis: Testing a model by developing a judging instrument. In S. G. Isaksen (Ed.), *Frontiers of creativity research: Beyond the basics* (pp. 367–389). Buffalo, NY: Brady.

© Springer Nature Singapore Pte Ltd. 2019
D. H. Cropley, *Homo Problematis Solvendis—Problem-solving Man*,
https://doi.org/10.1007/978-981-13-3101-5_2

   To show that this scale—the Creative Solution Diagnosis Scale (CSDS), see
Appendix A—has some reasonable degree of validity (i.e. does it really measure
what we claim it measures?), we have conducted several studies that ask people to
use the scale to rate the creativity of different products. Using a statistical procedure
known as factor analysis, we have been able to establish that these criteria hang
together in a sensible way, and that they are recognisable in a consistent way by
people using the scale. The technique also allowed us to weed out some unhelpful
indicators that weren't really helping our attempts to measure creativity, and end up
with a scale that can be used by anybody to rate the creativity of anything. School
teachers have used this scale to give their students feedback on the creativity of
assignments, and engineers have used it to rate the creativity or different artefacts.
I've even used it to rate the creativity of different designs of paper airplanes, in order
to illustrate the different qualities that contribute to a product's creativity.
   To help us understand the creativity of the inventions in this book, I have decided
to give each one a score, based on the preceding criteria. A product of any sort can be
measured on this scale—in other words, we can assign a number for each of the four
elements of creativity (effectiveness, novelty, etc.). A typical scheme is to score each
criterion on a scale from 0–4. Thus, a score of "0" means a complete absence of the
element in question. A "0" for either effectiveness or novelty would be fatal for our
purposes, and you won't be surprised to find none of these in the book. However, it
is possible that we'll find products that don't achieve 4/4 for everything. To the best
of my ability, I will score the products based on what would have been appropriate
at the time of their invention. Of course, with the passage of time, what may have
been highly original 1000 years ago may no longer seem so remarkable to us in
the twenty-first century. So the scores must be considered in context—how creative
would the product have been, at the time of its introduction? To help orientate our
thinking, I will also define the *total* creativity scores (out of 16) as follows: (a) 0–4 is
low—an invention scoring in this bracket overall is not really creative at all; (b) 5–8
is *medium*—scores in this range indicate an invention that is still rather uncreative;
(c) 9–12 is *high*—now we are starting to see inventions that are generally creative in
nature, and; (d) 13–16 is *very high*—these are impressively creative inventions!
   In education, there are two contrasting ways of viewing assessment of this type.
Norm-based assessment looks at individuals in comparison to each other. Under
this scheme, a class of students is expected to show a normal distribution of results
around some mean value. This means that only a small number of students in the
class can be expected to achieve a result far above average, and only a small number
far below average, with the majority scoring within a narrower band around the mean
value. By contrast, *criterion-based assessment* is concerned with the achievement
of set outcomes. Under this scheme, if you demonstrate that you can do what is
required—from tying your shoelaces to solving a differential equation—then you
pass, and if you cannot demonstrate the desired outcome, you fail. One of the features
of norm-based assessment that has always concerned me is that it is often used to
limit the number of high scores that can be awarded in a class—the idea of *grading
to the curve*. This seems to be inherently unfair if a given class has anything other
than a normal distribution of ability. Normal distributions are, of course, powerful

and useful, but they are based on the characteristics of populations, or representative samples of populations, and it is easy to see why this assumption might be incorrect in any given class of 28 children.

All of this, of course, is irrelevant to this book, except that I want to explain that I am taking a *criterion-based* approach to the assessment of the creativity of the inventions covered in this book. In other words, I judge each on its own merits, using the CSDS. I do not have a limit of one invention that can receive the maximum score, with the majority required to hover around some mean value. Not only is the list of innovations highly selective, and therefore in no way representative of the set of all things ever invented (in which case it might be possible to make some norm-based analysis), but the basic purpose of this analysis is that if something is creative—if it is relevant and effective, novel and so on—then it gets a high score, regardless of the creativity of anything else. Having said all that, it will still be interesting to see if there is any regular distribution of my scores, after the fact. The key point is that I am scoring them in a criterion-based sense. I am trying hard to be impartial and objective. Therefore, it is entirely possible that every innovation will have a score of 16/16, or 0/16, or there may be a normal-like distribution of the scores that I give. Let's wait and see!

Another important consideration regarding the creativity of products (solutions, artefacts, ideas) is the fact that novelty, in particular, is not a static quality. It is probably obvious that the moment an idea or a thing is revealed to the world, its newness begins, so to speak, to wear off. I often give talks about creativity, and the characteristics of creative solutions, and I use a paper airplane problem to illustrate this. I start by making a good, old-fashioned, paper dart. I then ask the audience to raise their hands if they have ever seen one of these—the paper dart—before. Everybody raises their hand. I then point out that if you've seen it before, can it really be considered novel?! One of the challenges that businesses face is how to introduce a new product to the market, the iPhone, for example, but at the same time, stop other companies from copying it. It's an impossible task because the moment you launch the product, you no longer control who knows about it. Not only that but the effectiveness and relevance of the product seem to be linked to the novelty. Once you launch the product, competitors jump in and bring out rival products, some of which may be better, and the effectiveness of your own product seems to decline. We talk about products becoming obsolete.

So, to a large extent, effectiveness and novelty are intertwined. The problem is, the only way to maintain novelty is to keep your product a secret—i.e. don't launch it! Obviously, that's a ridiculous proposition, so instead, companies accept this decline, and try, at the very least, to slow the decline of novelty (and the associated decline in effectiveness), for example, by keeping the product under wraps until immediately before it is launched. Even better, they accept this novelty decline as a fact of life, and work around it by launching a new, better product, fairly quickly and as soon as the novelty/effectiveness of the original starts to decline.

Interestingly, we don't just see this novelty/effectiveness relationship, and decline, in things like iPhones and other consumer products, and we didn't work out this relationship in the twenty-first century. As you will see in a later chapter, the Chinese went to great lengths to keep the invention of paper a secret from the rest of the world because they realised they had a valuable product on their hands and wanted to preserve their competitive advantage. Indeed, it took some 650 years before the invention of paper leaked out, first to the Middle East, and eventually, to Europe about 1000 years after it was invented.

Perhaps the most interesting and counter-intuitive example of the relationship between novelty and effectiveness is found in terrorism. This link first gave me an insight into what we called malevolent creativity, back in 2005. In the wake of the terrorist outrage of 9/11, I began to hypothesise that creativity was not something exclusively confined to nice people doing nice things, like engaging in business. It was clear, from the outset, that the 9/11 terrorists had done something very surprising, and very effective, and I started to explore the connection to the creativity of things.[4] What stood out was that over the space of only an hour or two, the terrorists' *product*—hijacking planes and flying them into buildings—went from highly novel and devastatingly effective, to not quite so novel, and not nearly as effective. The key was that the passengers of United 93—the people who fought back—had found out about the attacks, and decided to take action. In creativity terms, the terrorists' novelty had declined to a degree that was sufficient to drag down the effectiveness of their particular attempt to fly a plane into a building. It also illustrated that if the terrorists had thought to take away people's cell phones, the novelty would have remained high, and they probably would have achieved their intended objective.

It is time, however, to return to our core purpose. What are some of the varied problems that our ancestors have faced, across the centuries, and how did they use their capacity for creative problem solving to devise highly novel and effective, and frequently very elegant and paradigm-breaking solutions to these needs? Let's begin with our prehistoric ancestors and turn first to the African plains, and see how early modern humans fashioned tools out of stone.

---

[4]Cropley, D. H., Kaufman, J., Cropley, A. J. (2008). Malevolent Creativity: A Functional Model of Creativity in Terrorism and Crime, *Creativity Research Journal, Vol. 20, Issue 2* (April), pp. 105–115.

Prehistory: The Dawn of Invention (<2700 BCE)

Our first time period is both the easiest to define, but also the most challenging. The very nature of this epoch, especially the lack of written records, means that we have no choice but to engage in some intelligent guesswork when it comes to understanding the needs—the underpinning problems—that drove the inventions that we will consider.

This era takes us from the days of the earliest anatomically modern humans, surviving in small family groups on hostile African plains, through to some early civilisations, such as the Pharaonic Egyptians and the Minoans of Crete and the Aegean. It is tempting to suppose that the needs of greatest relevance in this period were the *basic* ones—how to find food, how to find water, how to keep warm and yet archaeological evidence tells us otherwise. Our earliest ancestors buried their dead, for example, often with elaborate care, suggesting that there were also more complex psychological or social needs that drove a search for solutions. Nevertheless, our predecessors could ill-afford to indulge these higher needs for too long, when there were far more pressing matters. Today, most of us take for granted where our next glass of water, or our next meal, will come from. Thousands of years ago, however, one of our ancestors must have decided that there was a better way to strip a carcass and prepare some food.

In this chapter, we will look first at what appears to be the earliest known tool used by humans—the *hand axe*. Once we have explored that, we will then examine how our forebears tackled the challenges associated with moving themselves and the things they valued over longer distances, as they began to branch out and explore their world. To do that, we will focus on watercraft propelled by *oars*. Finally, as we move into the beginnings of civilisation—larger groups of people living together and cooperating in their survival—we will see how ancient women and men began to communicate with each other and to record their communications in an early form of *writing*.

© Springer Nature Singapore Pte Ltd. 2019
D. H. Cropley, *Homo Problematis Solvendis—Problem-solving Man*,
https://doi.org/10.1007/978-981-13-3101-5_3

The Hand Axe (Date Unknown)

Man is greater than the tools he invents—American Proverb

Our first invention is a simple example of a *material handling system*. The modern sense of this phrase usually refers to things like forklift trucks, but in a more general, technological sense, we are talking about a solution that operates on, or transforms, physical things. The prehistoric hand axe was just such a solution: a tool intended to make basic tasks such as stripping a carcass for food, or cutting a piece of wood, much easier. The hand axe seems, therefore, to be a very good instance of a solution working at the very base of Maslow's hierarchy. Our early ancestors must have spent much, if not all, of their time focused on basic needs. How to find and process food; how to find water; how to keep warm; how to rest and the hand axe was a solution designed to help with all of these needs. Let's begin by taking a closer look at this invention.

What Was Invented?

The hand axe appears to be the oldest human stone tool for which there is clear archaeological evidence. Usually made from flint or chert,[1] the hand axe (see Fig. 1), was typically an almond-shaped, so-called *bifacial* flake, which would have been held with the wider base cupped in the palm, and the narrower point, or the sharper edge, used to scrape, cut and strike objects. We are fortunate that many examples of well-preserved hand axes have been found in archaeological sites, because this allows us to understand, very clearly, the properties and qualities of the tool, and even to experiment with it. Notwithstanding academic debates regarding the purpose and types of hand axes—many experts describe large, roughly shaped *blanks*, deliberately thinned *blanks*, specialised tools such as an adze, and blanks that served as a source of smaller items such as projectile points—it is not hard to imagine our early ancestors holding one of these devices in their hand, and performing routine chores a little more efficiently as a result.

Another important fact about hand axes that is clear, even now, is that they took some considerable skill to make. It is a nontrivial exercise to shape a flint into a convenient, hand-sized tool, and to create and maintain a usefully sharp edge. This tells us two things. First, the creation of hand axes was a very deliberate act. Second, the end use—the value that these tools added to their user—justified the enormous expenditure of time and effort.

The hand axe is also the earliest, concrete evidence that we have, that hints at our ancient forebears' problem-solving capacity. These objects required a deliberate and

---

[1] *Chert* is a broad class of fine-grained sedimentary rock composed of microcrystalline or cryptocrystalline silica. *Flint* generally refers to varieties of chert which occur in chalk and marly limestone formations (Wikipedia).

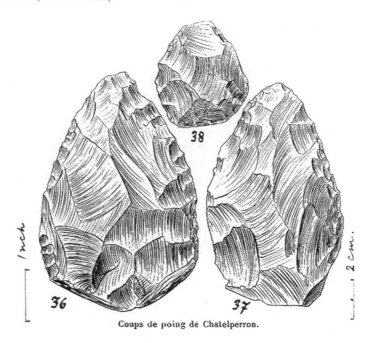

Coups de poing de Chatelperron.

Fig. 1 Palaeolithic hand axes (*Credit* Wellcome Collection)

conscious ability to *recognise* a suboptimal situation: a problem, in other words. Not only that, but it also necessitated the ability to imagine that there were alternative futures and the possibility of influencing these through their own actions. Once a problem could be imagined—that there could be an easier way to skin a carcass, or scrape an animal skin—our forebears then had to have the capacity to *generate alternative solutions*. The obvious solution may have been grabbing any object to hand, like an unshaped rock. However, these Stone Age humans also could see that some rocks worked better than others, so that they were able to *evaluate* different solutions, and notice what made them good or bad. Finally, they had the ability to recall what worked best, and also then to *design* objects with those desirable qualities. Perhaps most importantly, in exploring different alternatives, they didn't simply try the same unsuccessful options over and over, but when something didn't work, could adapt and try something new. It is this capacity—the ability to generate *new* solutions in response to failure, or changed circumstances—that defined our species' capacity for creativity.

Why Was the Hand Axe Invented?

It is impossible to know, with absolute certainty, what need or problem our Stone Age ancestors were seeking to satisfy with the hand axe, because, of course, there are no written records. However, thanks to careful archaeological research—for example, examining sites where both hand axes and other items have been found together—it is possible to speculate, with high degrees of certainty, that hand axes were used, for instance, to butcher animals. We can conclude this from scratch marks found on bones. In addition, it seems likely that this innovation was used to dig for edible roots, animals in burrows, and to uncover sources of water. It is also likely that it was also used to chop wood and remove tree bark, and even as a weapon. If we assume that hand axes were primarily *general purpose* tools, then what problem were our ancient ancestors trying to solve, for which a hand axe represents a good solution? The general nature of the solution demands a general purpose problem, and this may be *how to make everyday tasks easier*.

How Creative Was the Hand Axe?

Relevance and Effectiveness: The relative commonness of stone tools, and especially hand axes, suggests that these artefacts were widespread in prehistoric times. That fact then suggests that they were popular and that the information about how to make and use them was widely shared amongst people of the time. This further suggests that hand axes did the job they were intended to do. You could say that they were the claw hammer, or maybe the screwdriver, of their day. Perhaps an even better description is that they were the Swiss Army knife of the Stone Age. For this reason, I am confident in giving the hand axe a score of 4 out of 4 for relevance and effectiveness.

Novelty: the deliberateness with which hand axes must have been made—we know that flint knapping, the process of shaping flint and similar materials, was a considerable skill—suggests that some effort was invested in improving on the simpler, and more readily available, convenient rock. In other words, hand axes are a deliberate incremental improvement on whatever rocks were just lying around. Some early human must have thought "I can do better than this" when using a rough and unshaped rock to cut wood or strip an animal carcass for food. The design of the hand axe is a response to the more specific needs of cutting, scraping and so forth. It would have also been immediately obvious that some hand axes were particularly good ones, well suited to the task at hand. For this reason, there is a reasonable degree of novelty—3 out of 4 on our scale—represented by this innovation.

Elegance: As well made as many examples of stone tools are, I feel that we cannot ascribe too much elegance to this artefact. The craftsmanship of making a hand axe was nontrivial, however, it must still have been an uncomfortable tool to use for long periods. Its shape and design are obvious, but many small improvements are immediately apparent—protection for the user's fingers, a finer point, a sharper edge—all of which place a limit on just how elegant we can consider it. As a result, I give it a score of 2.5 out of 4.

Genesis: Does the hand axe fundamentally change—*would it* have fundamentally changed—the user's understanding of the problem at hand? I argue that it would not—it is simply an improvement on a basic solution of picking any convenient rock and bashing or scraping at bones, wood or animal skins. It is certainly better than an unshaped rock but did it fundamentally change the basic paradigm—no. Therefore I give it a score of 2 out of 4. It has some value in suggesting further improvements, but in itself, is not sufficiently radical in nature.

Total: The score of 11.5 out of 16 for the hand axe places it close to the top of the *high* range of creativity. It is still some way short of a *very high* score, and this can, as with many of our innovations, be attributed to both a weakness in elegance (i.e. imperfect execution), and a lack of disruptive qualities. It was a highly functional improvement over simple rocks, but it did not change the basic *tool* paradigm in the way that bronze or iron tools would do thousands of years later.

The Oar (c4000 BCE)

Destitutus ventis, remos adhibe—Roman Proverb[2]

The next innovation in our catalogue is the first example of an *energy handling system*. Unlike material handling systems, this class of inventions has the fundamental task of transforming energy from one form—e.g. the stored chemical energy in muscles—into another form, such as the kinetic energy of motion. Although Maslow's hierarchy of needs doesn't directly suggest a driving motivation for this innovation—it's not really a direct solution to any of the basic needs—it does directly *support* those basic needs. As our ancient forebears solved basic problems of food, water, warmth and rest, the ability to move more freely, for example by water, would no doubt have supplemented, or improved, those solutions. Boats allowed our ancestors to range more widely, looking for food, or trading goods needed to satisfy basic needs. Oars made that form of transport more efficient, therefore adding to the solutions applied to basic needs.

Of course, it could be argued that oars also helped to fulfil needs higher up the Maslow pyramid. Better, more efficient transport, might also have been important for satisfying psychological needs associated with friendship and belongingness, so that we see an innovation *directly solving* a particular problem, but adding something to other, peripheral problems. An oar converts energy—that's the real problem it solves. However, it also solves related problems associated with satisfying physiological needs. Even then, it contributes to psychological needs. This seems to be a good

---

[2]"If the wind will not serve, take to the oars". (Source: The Routledge Book of World Proverbs, J. R. Stone, 2006, Routledge, New York NY.).

illustration of genesis. Let's now look more closely at oars, and how they use simple principles of mechanical levers to propel boats through water.

What Was Invented?

Our earliest modern human ancestors were nomads: their survival depended on their ability to follow the animals they relied on for food, to seek out lands that had abundant and reliable sources of water and edible plants, and to occupy areas that provided adequate shelter and favourable living conditions. It's also not hard to imagine that early modern humans were always on the lookout for ways to ease the burden of being constantly on the move.

One of the first ways that humans appear to have tackled this problem was to use animals. However, I am going to treat that solution in the same way that I treat fire—it's a discovery rather than an invention. In other words, while horses, camels and elephants, for example, have been enormously important to human development over the centuries, their use as a means of transport—a better way of moving—is more a lucky coincidence than it is a deliberately *designed* solution.

Turning to our ancestors' inventive abilities, probably the first intentional solution to the problem of how to *move loads* was *the boat*. A boat, however, is nothing without a means of propulsion, but to appreciate why oars were developed, we first have to understand something about boats and how they were propelled before the appearance of oars. Watercraft (by which I mean something as simple as a dugout canoe or even a floating log) appear in the historical record at least as early as about 8000 BCE. Paddles—i.e. a stick with a wide blade at one end—appear at approximately 6500 BCE, and sails—simple structures designed to catch the wind and push a boat—appear as early as 6000 BCE. So we can see some sort of systematic progression in the development of watercraft, and increasingly sophisticated, and more efficient ways to propel them. Important for our discussion is the fact that oars were a relatively late development in watercraft propulsion. Oars emerged as a means for propelling watercraft sometime in the early to mid-Neolithic era. An early example, dating from between 5500 BCE and 4500 BCE, has been found in Hemudu, China amongst artefacts of the Zaoshi culture. A more recent example, believed to date from approximately 4000 BCE, was found in Japan's Ishikawa Prefecture. Oars, and rowing, appear to have been well-established in Egyptian culture by 1500 BCE, based on wall decorations and artwork in archaeological sites (see, for example, Fig. 2).

Although widespread in Europe and Asia for thousands of years, there is an interesting account of New Zealand Maori first encountering oars and rowers, during James Cook's first circumnavigation of New Zealand in 1769. They believed that Cook's men were Goblins, and part of the evidence they offered for this was that they must have eyes in the backs of their heads, because they rowed towards the beach where they landed facing away from the direction of travel. This confirms for

Fig. 2 Egyptian—wall fragment relief with men rowing (*Credit* Walters Art Museum, Public Domain)

us that oars were unknown as a form of propulsion to the Maori people, less than 250 years ago.

To appreciate how the oar functions, and how it represents an improvement over paddles and sails, we first need a quick refresher on levers. Class 1 levers have long been used for lifting heavy objects. A seesaw is a simple example, and this type of lever is characterised by the fact that the fulcrum—the pivot point—is situated between the resistance (the load to be lifted) and the applied force (often called the *effort*). The so-called *mechanical advantage*—the extent to which the lever makes the load feel lighter, or heavier, in a Class 1 lever can be less than, equal to, or greater than one, depending on the position of the fulcrum and the magnitude of the load and effort. It was a class 1 lever that Archimedes was referring to when he claimed that he could move the Earth.[3]

Class 3 levers, in contrast, are typified by tweezers, shovels and paddles. In this type, the applied effort is between the fulcrum and the resistance. For a paddle, this means that the top hand serves as the fulcrum, the bottom hand supplies the effort and the water is the resistance. Interestingly, the mechanical advantage in a Class 3 lever is *always less than one*, so its main benefit is as a so-called *speed multiplier*. In practical terms, this means that paddles are an inefficient way of moving a boat.

Class 2 levers, such as the oar, are characterised by the fact that the resistance is located between the fulcrum and the applied effort.[4] A wheelbarrow is also an

---

[3] "Give me a place to stand and I will move the Earth"—Archimedes understood that with a class 1 lever, and with the appropriate placement of the fulcrum, it was theoretically possible for him to lift the Earth. He recognised, however, that it would require an extremely long lever!

[4] A simple way to remember the different classes is with the mnemonic "FRE123". This describes the element—fulcrum, resistance (load), effort—that is between the other two, for a class 1, 2 and 3 lever. Therefore, the fulcrum (F) is between the resistance and effort in a class 1 lever, and so on.

Fig. 3 The oar as a class 2 lever

example of a Class 2 lever. In the case of the oar, the point at which the blade of the oar connects to the water is the fulcrum. The point where the oar connects to the boat (the *gate* in modern rowing boats) provides the resistance (in this case, the boat you are trying to propel), and the rower's hands provide the applied effort (see Fig. 3). Critically, the mechanical advantage of a Class 2 lever is *always greater than one*, and this class is therefore known as a *force multiplier*. This is why the oar is a more efficient means of propulsion than a paddle. Wind and sails, of course, required very little human effort, but as the Romans noted, when the wind wouldn't blow (or when it blew from the wrong direction) our ancestors had to rely on this efficient, manual mode of propulsion.

Why Was the Oar Invented?

In the twenty-first century, we ride a bicycle in preference to walking, or drive a car in preference to riding a bicycle, and it seems certain that our ancient ancestors were concerned with a simple problem: not just *how to move*, but how to move *efficiently*. Watercraft provided the means to move heavy loads, and the oar has proven a remarkably durable, efficient and reliable means for powering boats.

How Creative Was the Oar?

Relevance and Effectiveness: The fact that even ancient examples of oars have a similar shape and form to modern oars suggests that the basic concept is fit for purpose. The essential elements of the design of oars were worked out thousands of years ago, and they were able to satisfy the fundamental problem, namely, propelling a boat through water, using human power, in an efficient manner. For this reason, the oar scores the maximum of 4 out of 4 for this criterion.

Novelty: The impact of the first oars would have been immediately apparent to anyone using them. Put another way, it would be hard to imagine anyone in the ancient world reverting to paddling with their hands, or even a canoe-like paddle, after experiencing the efficiency and power of the oar. This innovation immediately makes the weaknesses of other human-powered mechanisms of boat propulsion obvious. The oar also reveals ideas for making it better—a lighter or longer oar, for example, would be better than a heavy, short oar. This puts the novelty of the oar at a fairly high level, and I have given it a score of 3.5 out of 4.

Elegance: The oar is also an innovation that seems highly intuitive. As engineers often say, a good solution usually *looks like* a good solution, and oars seem to satisfy this requirement. It seems obvious, just looking at oars, that they are designed to move a boat through water. This gracefulness, and sense of obviousness or completeness, gives the oar a fairly high score for elegance. Even ancient oars would have had a fair degree of aesthetic pleasingness about them—a straight pole or branch, stripped of leaves and other protrusions, with a wide blade slotted into one end and secured in place. Accordingly, I give the oar 3 out of 4 for elegance.

Genesis: Did the oar change our ancient ancestors' understanding of the basic problem (propelling boats)? To the extent that it applied mechanical aids to the basic problem, yes. To the extent that it also showed that there was more than one way to apply mechanical aids, and that not every mechanical aid had the same properties, again yes. Oars probably didn't turn the whole notion of propelling boats on its head, but it was an important, perhaps larger, step in a series of incremental improvements, and therefore gets a score of 2.5 out of 4 for genesis.

Total: The oar scores 13/16 in total, putting its creativity in "very high" category on our scale. Despite being invented thousands of years ago, it is interesting to see that oars were not found in some cultures, even as late as the mid-1700s. The Maori people of New Zealand, for example, had never before seen oars as a means of propulsion when Captain James Cook landed there in 1769. This is particularly interesting as it means that even an invention's longevity, as well as its apparent utility, is no guarantee that it will be universally adopted. Nevertheless, it scores well on our scale, and New Zealanders have long since made up for this late adoption of rowing by producing some of the best rowers in the world!

Cuneiform Writing (c2700 BCE)

Learned and unlearned, we all write—Roman Proverb

We turn now to the first example of an *information handling system*, namely a very early example of writing. Writing is interesting because it can satisfy very basic needs (e.g. security and safety, in the sense of communicating important information—"the enemy is coming!") but also serves as an example of *functional* creativity, placing it somewhat higher in Maslow's hierarchy. In fact, it can also satisfy psychological needs, and as a means of facilitating, for example, business and trade, is clearly an example of social creativity. In addition, writing also satisfies needs associated with both self-actualization, e.g. keeping a reflective journal, and special talents, e.g. through the medium of creative writing, and therefore is also a form of transcendental creativity! This immediately suggests to us that the history of human creativity is not a linear progression both over time, and from the base of Maslow's pyramid. In other words, humankind's creativity did not necessarily begin only with novel solutions to basic needs. Perhaps this simply shows how easy we are to please! The moment our ancient forebears were warm and had food in their bellies, they immediately started thinking about their next needs! This is actually important because it suggests that our ancestors were sufficiently adept at finding solutions to basic needs that they could *afford* the time and effort to solve higher order problems.

What Was Invented?

Like many of our topics, there is ongoing research and debate about exactly what constitutes the first example of the item/thing in question. Writing is no different, and recent discoveries, for example the Tărtăria tablets found in modern-day Romania, date from approximately 5500–5300 BCE. There is also the question of what was merely a means of recording numbers, as distinct from a means for communicating language. I will default here to what seems to be commonly agreed as the first representation of *spoken* language and discuss *cuneiform* (see Fig. 4) as it emerged in Mesopotamia in the period around 2700 BCE. Even if this can be debated, it doesn't detract from the basic purpose of the discussion—what need did early forms of writing satisfy, and was writing in general (or cuneiform specifically) a creative solution to that need/problem?

One further point of note—writing marks the transition from prehistory to history. Thus it is a nice way to finish our first epoch, where some speculation and guesswork is to be expected, and move onto firmer ground thanks to, well, writing!

Cuneiform was invented by the Sumerians, who occupied Mesopotamia (roughly speaking, modern-day Iraq) from approximately 4500 BCE until about 1900 BCE. This form of writing consisted of wedge-shaped marks pressed into clay tablets, using a reed as a form of stylus. The writing evolved from an earlier system of *pictograms*—icons that convey meaning through resemblance to physical objects—into a smaller set of characters that functioned as *logograms* (more abstract symbols that represent a word or phrase, similar to Chinese characters), *consonants* (think of modern, non-vowel letters) and *syllables* (units of pronunciation that contain a vowel—"wa-ter" consists of two syllables).

Some indication of the nature and utility of Cuneiform as a solution to a problem may be found in the fact that it became extinct by about the second-century CE. It gave rise to no *child* system of writing, and was therefore completely replaced by one or more systems that must have been of greater utility. The language itself remained extinct until the nineteenth century, when archaeological finds led to it being deciphered. It is estimated that as many as two million cuneiform tablets have been excavated, although only a fraction of these have been translated and published.

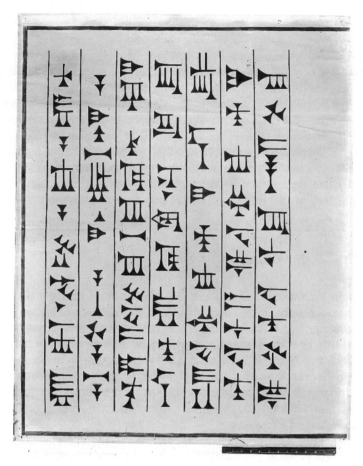

Fig. 4 Drawing of Cuneiform from the Ninevah expedition (*Credit* Wellcome Collection, Henry Layards)

Why Was Cuneiform Writing Invented?

Economic necessity appears to have been a strong catalyst for the development of writing, including Cuneiform. As early cultures in the Near East expanded and developed more sophisticated societies, the need for systems for sharing information, keeping records, recording commercial transactions, and so forth, emerged. At some point, the capacity of early humans simply to remember things was overwhelmed by the volume and complexity of the things that needed to be remembered. It's not hard to imagine a farmer on the banks of the ancient Tigris River trying to count and record how many sheaves of barley he had harvested—he probably scratched some marks with a thumbnail onto a piece of wood, much as any of us might do if we were caught without a pencil and paper. However, when a merchant needed to communicate this to another trader, in another town, this needed more than a notched stick.

Similarly, as early civilisations developed into more organised, and more sophisticated societies, how did these groups communicate the rules and standards needed to keep order and ensure the smooth functioning of society? King Hammurabi (1810 BCE–1750 BCE) ruled in the same region of Mesopotamia, and is known to us today for his set of laws—the Code of Hammurabi—which consisted of hundreds of laws dealing with contracts, wages, liabilities, divorce, reproduction and inheritance. A partial copy of this code was uncovered in 1901, carved into a 2.25 m high rectangular basalt slab, and used Cuneiform writing to capture Hammurabi's laws written in the Akkadian language. Although imperfect, and replaced by better, more functional languages, Cuneiform solved the problem of how to record language.

How Creative Was Cuneiform Writing?

Relevance and Effectiveness: Cuneiform, like other forms of writing that have developed throughout history, holds a special place in the pantheon of human creativity. As I have already noted, it is the distinguishing feature that separates prehistory from history, and without it, we would know very little of the innovations of earlier civilisations. Cuneiform, like all forms of writing, serves to capture and transmit language, and although the evidence—it became extinct—suggests that our forebears were able to improve on this early solution, the fact that we can read about the organisation of Babylonian society nearly 3700 years later means that I will give Cuneiform 3.5 out of 4 for relevance and effectiveness.

Novelty: It must have been apparent, even in its day, that cuneiform was an improvement on previous proto-languages. It was a new approach to an existing problem, with improvements and refinements that also opened up more possibilities for what writing could do, and how it could be used. At the same time, its eventual extinction suggests that there was still plenty of room for further improvement. For these reasons, I give this early form of writing 3.5 out of 4 for novelty.

Elegance: to understand the elegance of cuneiform we need to understand what it could do in terms of communicating ideas. Of all our inventions, this may be the hardest to judge dispassionately, not least because it is difficult for most of us to compare Cuneiform to our own system of writing. Unlike the hand axe, which we can easily imagine using, Cuneiform is so alien to us that it is hard to judge the extent to which it may, or may not, have been a well-executed solution to the problem of recording language. On the negative side of the ledger, it must have had significant weaknesses in terms of elegance, otherwise, why did it become extinct? One insight here is the difficulty that writers must have faced. Because Cuneiform was designed for impressing shapes on wet clay, it could not have been a form of writing that lent itself to jotting down notes and ideas. Even worse, when recording Cuneiform writing on a stone tablet, the process must have been slow and cumbersome. The same would be true for our modern alphabet, but the difference today is that we aren't restricted to writing on clay or stone. Given the limitations of supporting technologies, Cuneiform appears to have been a reasonably well-executed system of writing. The fact that modern scholars have been able to decipher it also suggests that, despite becoming extinct, it was still executed in a manner that made it possible to bring the information encoded back to life. For these reasons, I think we have to be generous to our forebears and give Cuneiform 3.5 out of 4 for elegance.

Genesis: We know, from the history and development of writing and language, that cuneiform was succeeded by incremental improvements, and eventually, by whole-sale changes. The weaknesses inherent in cuneiform helped to delineate subsequent improvements and establish the basis for the development of language and writing. Cuneiform also drew attention to previously unnoticed problems of writing, e.g. how to concisely capture all of the required vowel and consonant sounds of a language, or how to construct a set of symbols from which any words or concepts can be constructed. As soon as it became widespread, cuneiform also established a new benchmark for writing, against which other attempts could be judged. At the purest level, cuneiform also, to some degree, opened up new conceptualisations of whole problem of communication, and therefore had some degree of paradigm-breaking quality about it. I give it a score of 3.5 for genesis.

Total: Cuneiform scores an impressive 14 out of 16 for creativity. This places it well inside the very high range, and its score is evenly comprised of equal strength across all four criteria. There is no major weakness in this early form of writing, in terms of its fitness for purpose, its novelty, its aesthetic qualities and completeness, and its ability to frame and change the nature of the problem it was intended to solve. In each of these categories, there is some room for improvement—this is not surprising, given its eventual replacement—but it stands out close to the top of our scale for its innovative nature.

The Classical Period (753 BCE–476 CE): The Problems of Empires

The second epoch in our sequence, and the first that we can strongly associate with recorded history (thanks to *writing*), is the Classical Period of the ancient Greeks and Romans. Broadly speaking, this period spans about 1200 years, from the eighth century BCE (the founding of Rome took place in 753 BCE) up to the fifth century CE (the fall of the Western Roman Empire occurred in 476 CE). Of course, this leaves a gap of about 2000 years from the invention of the Cuneiform system of writing, however, I wanted to jump ahead to a period in which writing is well established, and for which there are clear, surviving written records. This does not mean that there were no interesting inventions in the period between 2700 BCE and 750 BCE! If we had more space, an interesting one would be the Code of Hammurabi, which I mentioned in the previous section on writing. This is one of the earliest known and documented legal codes and was proclaimed by the Babylonian King of the same name, who ruled in the region of Mesopotamia[1] from about 1791 BCE–1750 BCE. It was written on a stone column, using a Cuneiform script, and sets out a range of offences and associated penalties.

The Classical Period was a time of great advances in the Western hemisphere. Led by the ancient Greeks, many of the foundations of art, philosophy, democracy, science and education were first laid down and systematised, before being adopted, refined and spread across Europe and the Middle East by the Romans. Although *lost*, to some degree, after the fall of the Western Roman Empire, it was their *rediscovery* that would be a catalyst for the European Renaissance almost 1000 years later.

The first innovation that we consider in this period is Money—specifically *standardised coinage*—as a medium of exchange, tackling a core problem associated with trade and commerce. Our second invention in this era is the *construction crane*, developed in ancient Greece, and a vital advance in humankind's ability to manipulate energy for practical purposes. The final innovation that we consider in the Classical Period is the introduction of the *Julian calendar*, at the behest of the famous Roman ruler, Julius Caesar, in 45 BCE. This apparently simple information system would

---

[1]Roughly speaking, present-day Iraq.

© Springer Nature Singapore Pte Ltd. 2019
D. H. Cropley, *Homo Problematis Solvendis—Problem-solving Man*,
https://doi.org/10.1007/978-981-13-3101-5_4

play a crucial role in the organisation of society, of farming, and of many other key activities impacting on the daily lives of our ancestors.

Standardised Coinage (700–500 BCE)

Money won't create success, the freedom to make it will – Nelson Mandela, South African Statesman (1918-2013)

I finally know what distinguishes man from other beasts: financial worries – Jules Renard, French Author (1864-1910)

Money, as a distinct concept, has a surprisingly complex history. We live in an age of electronic transactions, and in a society in which it is increasingly possible to operate without actually carrying any physical money, either coins or banknotes. For many of us, money exists only as an abstract system for equating the worth of labour, goods, services and other things of value. Even the physical representation of money in the form of credit and debit cards is changing, with many people now conducting transactions solely on smartphones and computers. With the advent of decentralised, electronic cryptocurrencies such as Bitcoin, even the nature of money as a regulated, national system is changing.

The underpinning purpose of money, of course, has always been to represent or embody value and to facilitate trade. However, even before the invention of money in its more modern sense, humans were trading things of value, and for millennia money was, in a very real sense, an example of a material-handling system. It began, also, as a means of supporting the satisfaction of our basic needs—for example, simplifying the process of acquiring food—but over time, as we know, money has taken on greater significance in relation to our psychological needs. It's even possible to see that money plays a role in Maslow's highest form of needs, satisfying, or helping to satisfy, our need for self-actualisation and self-fulfilment. For some people, acquiring money is an end in itself. Regardless of these more recent aspects of money, let's now look at one of the earliest, recognisable forms of money, and consider coins—more precisely *standardised coinage*—as our next innovation, and the first in the Classical era.

What Was Invented?

Prior to the advent of written records, during prehistory, in other words, there is no doubt that early modern humans engaged in trade. Initially, this was trade in its simplest form. I give you some grain, for example, and you give me some animal skins. In these examples, the goods of interest are traded directly. At some point in our past, some civilisations also began to use other materials of value as a substitute for the goods that were traded. In this case, I might give you some attractive beads in exchange for your animal skins, and then trade those animal skins for more grain. This had certain advantages. I do not have to carry with me my grain, in order to make the trade with you. I can carry my beads as a more portable form of wealth. This system also made it possible and easier to have third parties involved in the process. In this case, an agent can operate on my behalf, using beads to conduct transactions on our behalf. However, the weakness of this system of bartering was that the value ascribed to the medium of trade (the beads) was not necessarily uniform. Clearly, it would be easier if there was some highly portable, but reliably and uniformly valued, medium through which the trade of goods and services could be conducted. This is the idea of *money of exchange*—physically tradable money such as coins or banknotes. Such a system also makes it possible to have *money of account*—a system of keeping records of the value of goods and service—if everybody subscribes to the same system.

Prior to the invention of standardised coinage, but long after the first systems of trading were established, humankind had started using precious metals as a more standardised system of money of exchange. This was further systematised in the period around 700 BCE–500 BCE, as the first manufactured coins appeared concurrently in India, China and cities around the Aegean Sea (Fig. 1 shows a coin from Athens in the fifth century BCE). Despite this advance, however, a problem soon emerged, namely, the question of guaranteeing that the amount of metal in the coin was as it should be. Initially, a *touchstone* was used to carry out this process of *assaying*, however, this process had inaccuracies and often required calculations to be made.

Touchstones were used as early as 3500 BCE in the Indus Valley, and in ancient Greece, as a means of testing the purity of gold and silver. Usually made from a stone such as slate or lydite, in its simplest form, a line was drawn on the touchstone using the metal in question, and the colour of the line compared to those made by metals of known purity. In this way, it was possible to estimate the purity of the sample in question. In more sophisticated forms, nitric acid was used to test the metal deposited on the touchstone. Pure gold (24 karats) would not react with the nitric acid. However, alloys, e.g. 22 karat gold (i.e. 22 parts gold, 2 parts copper), would react to some extent, thanks to the presence of the copper.

Standard coinage sought to resolve the issue of purity, with coins pre-weighed and pre-mixed (alloyed), and then stamped to indicate their origin. This meant that individuals could then be reasonably assured that a coin was worth what it was supposed to be worth. In fact, the problem that led to the development of standardised coins was not unique to money as a medium of exchange. The trading of spices was notorious, for example, in the Middle Ages, for cheating and dishonesty, with disreputable traders mixing other less valuable spices and substances in with their product, or wetting the spice to increase its apparent mass (see Cropley and Cropley, 2013).[2]

---

[2]Cropley, D. H., & Cropley, A. J. (2013). *Creativity and crime: A psychological approach.* Cambridge, UK: Cambridge University Press.

Fig. 1 Unknown coin (Tetradrachm) of Athens, 475–465 BCE (*Credit* J. Paul Getty Museum, Los Angeles)

Why Was Standardised Coinage Invented?

Standardised coinage was invented as a direct response to the problem of how to guarantee the value of the precious metals in trade, and therefore how to prevent fraudulent trading or cheating. If we were to state this as succinctly as possible (remember, in general, *how to verb noun* concept), we could say that the problem our forebears were trying to solve was "how to guarantee authenticity?" in the process of trade. Standardised coinage was the solution, so let's see how this stacks up as a creative solution to this problem.

How Creative Was Standardise Coinage?

Relevance and effectiveness: Standardised coinage scores 3 out of 4 for this criterion. Working backward from an assumption of full relevance and effectiveness, I have deducted a point here to reflect the fact that the early standardised coinage of the Classical Period was still vulnerable to fraud and cheating, despite the advancement that it had represented over the previous systems of touchstones and assaying. For this reason, while standardised coinage was a considerable incremental improvement, and while it fitted within clear constraints, it failed to eliminate fraud, and therefore did not completely achieve its purpose.

Novelty: Standardised coinage of this period certainly succeeded in drawing attention to the weaknesses of preceding systems of exchange. Gold, for example, could be mixed with small quantities of other metals in order to deceive the recipient, and the systems of touchstones and other assaying methods could not always detect these impurities, or were not always available. Even though the Greek philosopher, Archimedes, famously discovered the principle of displacement (the event that caused him to run through the streets shouting *Eureka!*) as a solution to the problem of detecting impurities in a crown that was given as a gift to King Heiro, methods such as this were less than fully effective. Standardised coinage, therefore, both demonstrated an improvement to previous solutions and also suggested how further refinements could improve the problem of detecting fraudulent trade. These characteristics should give standardised coinage a good score for novelty. However, while *incrementally* novel, standardised coinage did not represent a radical departure from previous solutions. Nor did this solution offer a fundamentally new perspective on the issue of the exchange of precious metals. People were still trading precious metal in exchange for other goods; the metal was simply more convenient, and it was to be hoped, more reliable and known state. Therefore, I give standardised coinage 2.5 out of 4 for novelty.

Elegance: The principal weakness of standardised coinage, with respect to elegance, stems from a mild deficiency in terms of the quality of execution, finish and proportion. While the first coins developed in the Classical Period are undoubtedly an improvement on previous media of exchange, I suspect their adoption and acceptance would have been slow. Imagine a trader receiving one of these new coins for the first time. Even with an official stamp, their suspicions must have been aroused. Is this coin really worth what is claimed? Can I be sure that it hasn't been tampered with in some way? It seems highly likely that there was a long period of transition, during which many traders must have resorted to old methods of assaying in order to be certain that these new coins were genuine. This hypothesised transition phase, during which trust in the new system had to be built, detracts from the elegance of the solution, and results in a score of 2.5 out of 4 in this criterion.

Genesis: Standardised coins do rather better against the criterion of genesis. This solution has, at its heart a new, paradigm-nudging foundation. Trade had commenced with the exchange of the actual goods of interest—I give you a chicken, and you give me some grain, for example. This developed into systems whereby some more convenient substance—a precious metal, for example—could be exchanged in lieu of the actual goods. This must have been particularly convenient for trade that involved intermediate, i.e. third parties. As we have discussed, however, both the goods themselves and precious metals could be tampered with, so that the recipient might not receive the true value of their trade. Standardised coinage was the next stage of this paradigm, and a particular feature, with respect to genesis, is that this suggests that anything might be traded, provided its value and provenance can be guaranteed. Even, of course, pieces of paper! The solution in question also has elements of genesis in how it offers ideas for solving other problems in which an exchange of anything—information, for example—takes place. We still use this concept in some IT security systems, in which a token—in effect, a digital coin—is exchanged as a means of establishing the security of a digital transaction. At the same time, standardised coinage also drew attention to previously unnoticed problems, such as the means by which the origin and purity of any medium of exchange can be guaranteed, and also the challenges of preventing unauthorised copies of the medium. As a result, standardised coinage scores 3 out of 4 against the criterion of genesis.

Total: The standardised coinage of the Classical Period, introduced by civilisations such as the ancient Greeks, has a total creativity score of 11 out of 16. This places it in the *high* range, with some particular weaknesses in novelty and elegance, reflecting first, a rather incremental approach to the problem of trade and the use of a medium of exchange, and second, the relatively crude nature of the coins. At the same time, standardised coinage opened up some new perspectives on the general problem of the secure exchange of many things—not just goods, but information, for example. Consequently, while a low score in our pantheon of inventions, we must remember that 11 is not *un*creative, rather it is simply a little lower than many of the solutions that we consider in this book.

The Construction Crane (c550 BCE)

Archimedes had stated that given any force, any given weight might be moved – Plutarch, Greek Biographer (45 CE-127 CE) in, Life of Marcellus

The construction crane, as it was invented by the ancient Greeks, is by no means our first energy-handling system. However, it is one of the earliest that began to draw on more complex principles of mechanics, and to combine more than one mechanical concept into a single device. The crane has very simple purpose, and what makes it an energy-handling system is that it transforms the chemical energy of human beings into the potential energy of an object raised against the force of gravity. One way to think about that is rather like a battery. We, generally, think of batteries as devices to store electrical energy, for later use, but the concept applies to other forms of energy as well. Indeed, we are beginning to see modern systems storing energy by pumping water to the top of a tower, for example. The energy can later be released, as electricity, by allowing the water to flow down, under the influence of gravity, and through a turbine that spins and generates electricity. So it is with large stone blocks. Human energy input, via a crane, raises a heavy block by giving it potential energy. If we want to release that energy, we let the block fall. In this way, the construction crane gave our ancestors one of the first tools for using, storing and transforming energy.

What Was Invented?

First developed by the ancient Greeks in approximately 550 BCE was the first construction cranes that utilised winches, pulleys and ropes to move heavy objects, such as stone building blocks, in both vertical and horizontal planes. In other words, these cranes could both lift an object and also move it forward or backward to some degree. Not in the sense we are used to do with modern cranes, but in a way that began to change both what could be constructed and how it was constructed in the ancient world.

The key elements of the first construction cranes were probably the *winch* (dating from around 500 BCE)—a simple mechanical device for winding up or letting out a rope under tension—and the *pulley* ($\approx$750 BCE)—a wheel designed to support and change the direction of a rope or wire under tension. Together, as a crane capable of lifting and shifting heavy loads, these gave rise to a device similar to the one represented in Fig. 2. The example shown is a Roman *Trispastos* crane and it is estimated that a single person could lift a load of approximately 150 kg using this device. This is based on two factors. First, the so-called *mechanical advantage* of the pulley system. This describes the ratio of the load that is being lifted, to the effort required to lift it. In very simple terms, the more pulleys joined together, the greater the mechanical advantage (but the shorter the distance the load is lifted). The second factor determining how heavy a load could be lifted by this crane is the effort that a person is capable of sustaining. If a single operator was pulling on the rope, and a load of 150 kg was attached to the crane, it would feel like 50 kg to the operator. The question then is how long a person could continue to exert that effort. Perhaps a better way to think of the crane was that a single person could do the work of three.

Prior to the invention of the crane, the typical method for moving heavy loads—e.g. the building blocks used to construct a pyramid—was to build ramps, and use large numbers of workers to drag them, perhaps with the help of rollers. The key point is that this method was very labour intensive. With the advent of the construction crane, not only could building materials be moved by fewer people, but buildings also began to use smaller blocks that could be lifted into place by the available cranes.

The construction crane is an invention that is really built around two key contributing devices—the pulley and the winch—and with their invention, and integration into a *system*, the crane changed what could be constructed. Even modern construction cranes differ only really in the lifting power that can be applied to the basic system of winch and pulley. So, what was the underpinning problem that our ancestors solved with this device?

Why Was the Construction Crane Invented?

My first instinct is to focus on basic needs, in the sense of Maslow. A civilisation that could build bigger and stronger structures gained advantages in terms of shelter, rest,

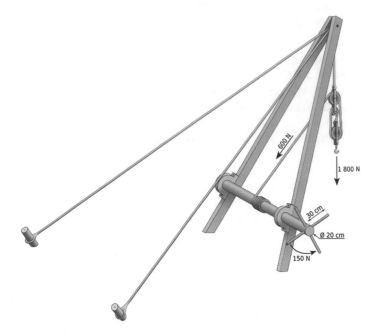

Fig. 2 Roman Trispastos Crane (*Credit* Wikimedia, Eric Gaba)

safety and security. The crane was simply a more efficient way of satisfying those basic needs. The weakness of this argument is that the most enduring evidence of past civilisations like the Greeks and the Romans is the buildings that they constructed. The Parthenon in Athens was begun in 447 BCE and completed in 432 BCE, and must have been constructed with the aid of construction cranes. Similarly, the Colosseum in Rome, built in the period 70–80 CE, benefited from the use of cranes. Neither structure was built to satisfy a basic need such as shelter; they were monuments to their civilisations, demonstrating and celebrating their wealth, power and influence. In this sense, the crane may be much more a solution to higher order needs such as how to demonstrate wealth or power. Regardless of these loftier purposes, it seems certain that builders, faced with the day-to-day task of creating a structure, found their job was easier with the invention of the construction crane—a device designed to solve the problem of how to lift heavy objects.

How Creative Was the Construction Crane?

Relevance and effectiveness: The first construction cranes, dating from at least 550 BCE (there are structures from ancient Greece still standing that predate this time, and which almost certainly would have needed cranes in order to be built), were effective at solving the key problem of lifting heavy objects such as stone blocks.

Although crude by modern standards, and requiring human power to operate them, the construction cranes of ancient Greece nevertheless were highly relevant to the problem, with only one deficiency related to their performance. Clearly, the design, the materials used to construct the cranes and the limitations of human power meant that the crane's lifting capacity was rather limited. Although this capacity was far beyond what was possible without this device, this must limit the performance, and consequently, I give the construction crane a score (3.5) that is slightly below the maximum for relevance and effectiveness.

Novelty: The first construction cranes of ancient Greece were a considerable improvement on the preceding methods that relied almost exclusively on raw human power, possibly supplemented by the simplest of mechanical aids such as rudimentary levers. The improved capacity of the crane, utilising both pulleys and winches, immediately draws attention to the limitations of the simple tools and methods of construction that preceded it. The cranes must also have caused builders to quickly recognise the potential to extend and augment the same basic principles even further. However, the construction crane cannot be said to be a radically new approach to the problem of lifting heavy objects—it was a blend of incremental improvements to existing devices (pulleys and winches), combined in a new and useful way. This detracts a little from this innovation's novelty, but still leaves us with a score of 3 out of 4.

Elegance: The evidence that is available for the design of the first cranes in ancient Greece suggests a reasonably well-executed design. To a modern eye, the structure might appear rather crude, but we can imagine that for people present at that time, the construction crane must have seemed like a clever and highly pleasing solution. It made the process of lifting heavy stone blocks used in the construction of monumental buildings far simpler than had been possible previously, and it is not difficult for people to understand and appreciate the basic mechanical principles that allow a pulley system to lift heavy weights. No doubt, there was room for improvement in the execution of the device, and many ancient cranes must have been rather dangerous devices to operate. However, we can still give this innovation a healthy score of 3 out of 4 for elegance.

Genesis: The first construction crane scores 3 out of 4 for the characteristic of genesis. It has all the requisite elements to a strong, but not maximum, degree. For example, these devices set the observer on a path of imagining better and newer ways to use basic mechanical aids to form new and useful systems. The crane also had elements of transferability—solving apparently unrelated problems—and we can see this in the application of the same basic device as a weapon. The *Claw of Archimedes* was a crane-like device intended to lift enemy ships into the air and drop them, thus smashing them. The crane also set a new standard for judging mechanical aids for lifting, pulling and other heavy tasks beyond the strength of unaided humans.

Total: The first construction crane scores a total of 12.5 out of 16 for creativity, and this places it between the high and very high ranges. Like a number of other inventions in this space between the two categories, the crane has a great deal to admire in terms of innovation, without achieving a maximum score in any particular category. It was a *pretty good*, but not a perfect solution to the problem of lifting heavy objects. It was quite a novel approach, especially in how it combined existing technologies. It was quite well executed, and it pushed the boundaries of the current paradigm, without quite breaking them.

The Julian Calendar (45 BCE)

The calendar was a mathematical progression with arbitrary surprises – Paul Scott, English Author (1920-1978) in, Towers of Silence

The next invention in our catalogue is an example of an information-handling system. Unlike energy or material-handling systems, it did not have an obvious physical effect, and yet, like information-handling systems that would follow it, the impact may be just as profound. It is tempting to assume that this invention—the Julian calendar—must also have no real role in satisfying humankind's basic needs. How could an information-handling system help to satisfy our need for food, water or rest? However, as we will see, it is just as central to satisfying those basic needs as a hand axe, and possibly more so. If physical tools were necessary for harvesting a crop, and providing food, calendars helped to make sure that there was a crop to be harvested. The Julian calendar was not the first attempt by our ancestors to improve activities such as farming through better information, but it was a significant improvement, and it helped to make the satisfaction of basic needs more efficient, more reliable and more predictable. It was also given to us by one of ancient history's more remarkable characters. We probably think of Gaius Julius Caesar (Gaius was his *praenomen*—his first name; Julius was his *paternomen*—his family name; Caesar was his *cognomen*—his nickname) mainly for his political and military accomplishments. However, he was also a renowned orator, poet, author, grammarian, as well as inventor of calendars. Let us now take a look at his invention.

What Was Invented?

Modern humans first lived nomadic, hunter-gatherer lives. Prehistoric technology, as we have seen, consisted of stone tools, animal skins and other rudimentary artefacts that simplified the process of living off the land in small, self-sufficient groups. Even in these primitive times, our ancestors made use of simple measures of time. The transition from day to night, to day again, gave modern humans a reliable, ever-present means of marking the passage of time. From this, we can speculate that they could share related information—how long it takes to walk to a known source of water or food, for example. In this way, it gave them a degree of control and order over their lives. We also know that prehistoric civilisations understood the 28-day cycle of the moon (the lunar month), and even the regularity of the sun's orbit around the earth, manifest as seasons. There were two weaknesses in this system, however, that probably made little difference to nomadic cultures, but would fail to meet the needs of larger, more static, agrarian societies that emerged from the last Ice Age. First, these primitive calendars were inherently *inaccurate*, and second, they did not lend themselves to making reliable forecasts about future events.

As the population of modern humans expanded, a need emerged for more reliable and more plentiful sources of food. Farming emerged from this change, and with it, a need for more accurate measures of time—better calendars, in other words (Fig. 3).

In both Mesopotamia and Egypt, early calendars improved on the direct observation of day/night and lunar month by building a more sophisticated observation of celestial bodies. This enabled those with the requisite specialist knowledge not only to mark the passage of time in days and months but also more accurately to measure the passage of the solar year, and predict the changes of seasons—both of vital importance for farming. We know, for example, that the ancient Egyptians depended, for their farming practices, on the regular flooding of the Nile River, and had the ability to predict this event. A year measured on a lunar calendar is only 354 days, and would have made these predictions extremely difficult without the added sophistication of other celestial observations.

The ancient Egyptians used observations of the star Sirius to make their accurate predictions of the Nile's annual flood. They improved on the inaccuracy of the lunar calendar through this means, understanding that the solar year is 365 days long. This improved the inaccuracy of the lunar calendar from a 32-year cycle—the time it took

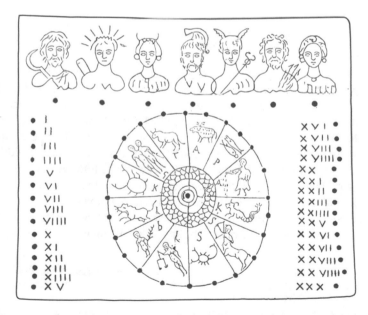

Fig. 3 Roman calendar—parapegma (third–fourth century CE) (*Credit* Wikimedia, Лобачев Владимир)

for the lunar and solar calendars to resynchronise—to a cycle of 1460 years.[3] Much better, but still with a drift that would eventually appear to make summer occur when it was supposed to be winter, and vice versa.

The Egyptians and the Romans who emerged at a later date both failed to understand that the real length of the solar year—the actual time it takes the Earth to complete one full orbit around the Sun—is 365 days, 5 h, 48 min and 46 s, i.e. not quite 365.25 days long. By the time of Julius Caesar—in the first century BCE—the existing Roman calendar, based on the lunar month, was a mess. It is reckoned that in Julius Caesar's time the calendar disagreed with the actual seasons by about 3 months—in other words, although the calendar said June 1 (the start of Summer), the weather conditions said early Spring (or early Autumn). Clearly, not only unhelpful for time-dependent activities such as farming, but also a hindrance to the efficient operations of a sophisticated, militaristic and expanding society such as Rome.

Julius Caesar sought to fix this problem with a new, better calendar. His calendar, introduced by edict on the first of January, 45 BCE, used a pattern of 3 years of 365 days, followed by one leap year of 366 days, to give an average year length of 365.25 days. Sounds familiar, and it seems as though this should be perfect. The slight catch is that the real solar year is 11 min and 14 s less than 365.25 days. In other words, the Julian calendar still contains a very small error. This translates into a gain of one day every 128 years. Not bad, but something that had to be corrected eventually,[4] as I'll describe shortly.

Why Was the Julian Calendar Invented?

One way to express any problem, succinctly and clearly, is in the form of a "how to verb noun" statement. We do this in engineering, and it is a technique that is very helpful in stimulating the search for creative solutions. The more general (or abstract) you make the verb and the noun, the easier it is to find a range of different solutions to the same problem. There is a small catch. Frequently, in our problem-solving processes, we need to capture both WHAT the solution has to do (the "how to verb noun") but also HOW WELL it needs to do it. The how well is often critical in solving a problem, because it puts some boundaries on the solution. If I ask you to solve the problem of how to get to work, it's hard to come up with a satisfactory solution without knowing "how well". In other words, the how well tells us whether you need to get to work in 30 s, or 1 h, and that will dictate what solutions are

---

[3]In other words, the lunar calendar would quickly fall out of synchronisation with the actual solar year, but with the Egyptian's better system of marking time, this drift out of synchronisation was much slower.

[4]You may be wondering how the Gregorian calendar avoids the same problem, given that we employ the same basic leap-year process. That seems to suggest that the Gregorian calendar also assumes an average year length of 365.25 days. The fix is that we also have some leap years removed, in so-called *centurial* years that are not divisible by 4. Thus, 1700, 1800 and 1900 were NOT leap years, but 2000 was.

acceptable. I think that the same concept applies in the case of the Julian calendar. The problem is something like how to measure time. However, it really needs a how well condition as well, otherwise we risk ending up with a solution that was just as bad as the lunar calendar. So I will define the problem that the Julian calendar solved as "how to measure time accurately", as a better definition of the problem.

How Creative Was the Julian Calendar?

Relevance and effectiveness: The Julian calendar is one of the few examples that I am not going to give a score of 4 for this criterion. While it was a major improvement over previous calendars, it still had some minor flaws that would only be fixed with the introduction of the Gregorian calendar, in 1582.[5] The Julian calendar was good, but not quite perfect, and therefore as a true solution to the problem of measuring time accurately, I can only give it a 3.5 out of 4.

Novelty: The Julian calendar represented a significant improvement over previous calendars. It kept the innovation process moving along the curve of incrementation, adding an improved outcome without passing a point of diminishing returns. It highlighted the weaknesses of previous calendars and showed, decisively, how these could be improved. However, it did not radically change the calendar paradigm, and some continued weaknesses meant that there remained value to be squeezed from the concept. I score the Julian calendar a 3 out of 4 for novelty.

Elegance: While it gives a relatively skilfully executed solution, which was logical, consistent and self-evident, the previously identified weakness of the Julian calendar leaves a slight sense of incompleteness that must have been evident even in its day. The Romans understood that their previous calendars were out of synch with reality, and their goal was to rectify this. However, they must have suspected that the new calendar was not quite complete—it contained a small, but ultimately significant error. This detracts a little from the elegance of the Julian calendar, leaving it with a score of 3 out of 4.

Genesis: Did the Julian calendar changes how the problem of measuring time was understood? Not really. It was, as I have said several times, a good improvement on an existing concept. It was so good that it worked quite well for over a 1000 years. However, of all our innovations, this was perhaps one of the most purely incremental, and it is hard to find much that is truly disruptive about the Julian calendar. Maybe that reflects a general characteristic of the Romans—they were nothing if not practical, and very good at taking existing solutions and making them better. I give the Julian calendar a score of 2.5 out of 4. "Et tu, Dave!"[6]

---

[5]The Gregorian calendar is the one we still use.

[6]"Even you, David!" In Shakespeare's play, *Julius Caesar*, our hero utters "Et tu, Brute!" despairingly as he recognises his friend, Brutus, among his assassins. Sorry Gaius, but it's a 2.5 for genesis, and that's that.

Total: The Julian calendar gets an overall score of 12/16 for creativity. This places it in the "high" category, but is let down by a slight weakness in relevance and effectiveness—it didn't quite solve the problem—and by a lack of any real paradigm-shifting qualities. Nevertheless, in many ways, the world runs on incremental solutions—the better, faster and cheaper improvements that we make every day, with occasional bursts of radical innovation sprinkled around them. Indeed, many companies operate on a similar basis. The 70/20/10 model is quite common—what Coca Cola calls the Now, Next and New. The idea is that businesses should spend around 70% of their effort on their current products, about 20% of their effort on what they will do next (the incremental solutions) and about 10% on the new, radical, things that will turn into their core business further down the track. The Julian calendar was vital in its day, and persisted for 100s of years, but was slightly imperfect.

The Dark Ages (476–1453 CE): Eastern Creativity

Next in our sequence of time periods is the span of approximately 1000 years known (in Western culture) as the Dark Ages. This epoch spanned the fifth to fifteenth centuries of the common era, beginning approximately with the sacking of Rome—and the fall of the Western Roman Empire—in 476 CE, and ending with the fall of the Eastern Roman Empire[1] centred on Constantinople (modern-day Istanbul), in 1453 CE.

The very name of this period—the Dark Ages—suggests that we might not find much in the way of human creativity and innovation. However, while some of the advances, organisation and sophistication of the Western world evaporated with the sacking of Rome, neither the world in general, nor even Europe, stood still. Many important inventions were created across this time period, even in Europe—the Quill Pen, Spectacles and the Coiled Spring, to name just three. However, if we leave Europe temporarily to the chaos of the marauding Ostrogoths, Vandals and other barbarians, we can turn our attention Eastwards at this time, and we find humankind as creative as ever.

Our first innovation in this new period, and one which would take two of our preceding inventions—writing and the calendar—to new heights, was *Paper*. This Chinese invention would take some time to make its way to Europe, via the Middle East, and has been critical to the communication needs of humankind over a period of 2000 years. The second innovation that we consider in this epoch is the development of the *Spinning Wheel*. A very simple invention—really just a way of improving what our forebears did quite well unaided—but with profound consequences in terms of improving the lives of people, making better use of their time and stimulating what people could do productively with their time and abilities. Our final invention for the period of the Dark Ages is the invention of the humble *crankshaft*. This mechanical device transformed motion between rotational and reciprocating forms—something that sounds uninspiring—but without which the advances seen 500 years later during the Industrial Revolution would not have been possible.

---

[1] Also known as the Byzantine Empire, or Byzantium.

© Springer Nature Singapore Pte Ltd. 2019
D. H. Cropley, *Homo Problematis Solvendis—Problem-solving Man*,
https://doi.org/10.1007/978-981-13-3101-5_5

Paper (105 CE)

The visionary starts with a clean sheet of paper, and re-imagines the world – Malcolm
Gladwell, Canadian Author (1963–)

Paper is thought to have been invented in the year 105 CE by a Chinese courtier
(and eunuch)—Ts'ai Lun—in the court of Emperor He of the Han Dynasty. You may
have noticed that this date falls outside of the definition of the Dark Ages, and is,
strictly speaking, in the Classical Period. However, I include it here to highlight a
contrast between the Eastern and Western worlds, and indeed, the place of the Arab
world between these. Although paper was invented by the Chinese in 105 CE, it did
not reach Europe until the early 1100s CE (firmly in the Dark Ages), and did so via
the Middle East.

Although we could say that paper represents a material-handling system, I make
the case here that it is an important example of an *information*-handling system. This
also reflects the kinds of needs that paper has helped to satisfy over many centuries.
As is the case with many of our inventions, there is frequently an *obvious* need that is
satisfied, but a number of other secondary or related needs that are also satisfied by
the innovation in question. Paper could be said to help satisfy basic needs—e.g. food,
as means of facilitating trade and food production—but it also clearly plays a role in
satisfying higher order needs, such as those associated with love and esteem. How
else could our ancestors have poured out their feelings to a loved one, or maintained
relationships with distant family members, except through the information-handling
qualities of paper? Finally, through some of humankind's most sublime works of
artistic production—poetry, literature, music and art—our needs for self-fulfilment
and self-actualisation have been well served by paper.

What Was Invented?

Ts'ai Lun's invention was a major refinement of more traditional forms of writing
surfaces—e.g. papyrus, bamboo, silk and wood—and consisted of both the new
materials used to make the paper, and a new process for making this novel writing
material.

It is very interesting to note that while the modern process of making paper has
improved significantly, through the application of more modern tools and technology,
the underlying process is essentially that same as it was in Ts'ai Lun's time. What
we now regard as paper is made by forming thin layers of fibrous materials (in Ts'ai
Lun's day, tree fibres, wheat stalks and the bark of Mulberry trees; today, the cellulose
derived from wood pulp—see Fig. 1) suspended in water, then draining the water
and drying the resulting thin sheets.

The Chinese clearly attached significant value to this invention and tried to keep it
a secret. However, after the Battle of Talas, in 751 CE—fought between an alliance of
Arab and Tibetan forces, against soldiers of the Chinese Tang dynasty, in the region

Fig. 1 Example of early Chinese papermaking (*Credit* Wikimedia)

of the modern-day border between Kazakhstan and Kyrgyzstan—the Arab forces captured Chinese paper merchants and took this technology back to the Middle East. Modern paper then made its way into Europe, via Moorish Spain in the twelfth century CE. Even then, it took some centuries before Ts'ai Lun's paper replaced parchment (a writing surface made from animal skins) and later, paper made from linen and cotton rags.

Why Was Paper Invented?

The problem that underpins the invention of paper is one that traces its origins back to one of humankind's earliest challenges—how to communicate? As human's learned to communicate through language, and as civilisations became more sophisticated and complex, the detail and volume of what needed to be communicated increased. We have already seen how the invention of writing addressed a similar need at the end of the prehistoric period. It seems that the better we became at communicating through written language, the more useful this became, and the more there was a need to make this information easily accessible and shareable. It's rather like the situation with modern computers. Without the means to store information (in computer memory) or share it (via networks), computers themselves have a limited usefulness. Paper, it seems, functioned in the same way as memory and networks, allowing people to connect and communicate more easily.

How Creative Was Paper?

Relevance and Effectiveness: Ts'ai Lun's paper fares well in any assessment of relevance and effectiveness. It drew on existing knowledge of natural fibres, and writing surfaces, to deliver an improved solution that fulfilled its intended purpose. Very simply, it solved the problem of providing an affordable, versatile writing surface that would facilitate the storage and communication of information. Not only that but insofar as it eventually replaced other common writing surfaces—parchment, papyrus, silk—across the world, it clearly demonstrates its effectiveness. This means that it must receive the maximum score for relevance and effectiveness, namely 4.

Novelty: Paper is an interesting example of a blend of incremental and radical innovation. In China, where it was developed, it replaced substances such as bamboo and wood (used in their unprocessed forms, which made for heavy, unwieldy writing surfaces) and silk (which was expensive, and therefore of limited availability). It redirected existing solutions towards a perfect compromise—the cost-effectiveness and availability of wood or bamboo, and the thinness and utility of silk. Furthermore, it did this through an approach that was substantially new and different—in effect, extracting the key ingredient (cellulose) from plant matter, but doing away with the bulky fibres. Some people suggest that Ts'ai Lun got this idea by observing wasps and their nests. Whether that is true or not, at some level, he realised that plants, such as trees, contained something that could be used productively to create a writing surface. I give paper a score of 3.5 out of 4 for novelty.

Elegance: This quality can often be associated with aesthetic appeal, and there is some logic in associating "looking nice" with "working well". However, elegance is more than just the visual appeal of a solution. Or perhaps it is more correct to say that elegance seeks to capture what we are really assessing when we say that something looks nice. One of these qualities is that elegant solutions have a simplicity and obviousness about them. Typically, you can immediately see that the solution, whatever it is, fits the problem. I argue that this is the case with paper. If we imagine ourselves back in China in the Han dynasty, tired of writing on pieces of bamboo or wood, or unable to get sheets of silk, a sheet of Ts'ai Lun's paper must have made an immediate impact on the observer. We could take the same writing implements we used for the other surfaces, and begin writing on this new surface. Not only that, but it could be folded, strung together, carried easily and so on, all giving it a neatness—an elegance—that was instantly appealing. Of course, the early examples were probably imperfect, with some room for refinement, but I give the first examples of paper a fairly high 3 out of 4 for this quality.

Genesis: Wouldn't it be nice if everything in life was simple and straightforward? Here's the dilemma—paper has turned out to be enormously important as a component of a wider system of spreading knowledge (e.g. in concert with the printing press), and ultimately, changing the world. And yet, in 105 CE, while it was new and effective, and a neat solution to the writing problem, we risk overstating its level of disruption and paradigm-changing qualities. Indeed, paper would not have this wider impact for another 1000+ years, and only then because both paper and printing presses were a means of capturing and sharing ideas that depended on the invention of writing, thousands of years before. Put that another way—the impact of the printing press was a function of about 4000 years of progressive problem-solving. Paper certainly had an impact in changing how people understood the problem associated with capturing and sharing their written ideas. It set a new standard for judging solutions to this problem. However, its proximal impact—i.e. its local, near-term impact was more incremental than radical, so that I feel I can only give it a score of 2.5 out of 4 for genesis. However, I can see what's coming—we may need to consider some, if not all, of these inventions through a modern lens as well. In other words, once we have been through our time periods, it may be valuable to look back at each invention from a twenty-first-century standpoint and analyse the creativity *ex-post-facto*,[2] or after the fact. Is an invention that was creative in its day still just as creative now or has that creativity declined or increased with the passage of time?

Total: The preceding comments notwithstanding, paper fares well on our scale of creativity. The score of 13 out of 16 place it in the *very high* range, with minor weaknesses in elegance and genesis. It was certainly highly functional, and is one invention that has genuinely stood the test of time.

The Spinning Wheel (c700 CE)

We sleep, but the loom of life never stops, and the pattern which was weaving when the sun went down is weaving when it comes up in the morning. Henry Ward Beecher, American Social Reformer (1813–1887)

Necessity, as readers will remember, is the mother of invention, and basic needs have a way of imposing an urgency that frequently can't be ignored. When we can't breathe, there is nothing more urgent than finding a lungful of clean air. As the most urgent need is satisfied, we turn our attention to the next most important, moving from breathable air, to water, to food and so on. We can manage for only 3 minutes without air, 3 days without water and 3 weeks without food. Somewhere in this

---

[2]Latin—literally, *after the fact*.

mix, and depending on where our ancestors lived, warmth and protection from the elements have always featured strongly. Clothing has, therefore, always been an important concern for humankind, particularly as our ancient ancestors moved out of Africa, and into higher latitudes. Animal skins provided one solution, but our ancestors soon learned to use natural plant and animal fibres as a source material for making better and more functional clothing. The Spinning Wheel is a material-handling system. Its purpose was to facilitate a physical transformation of thin, weak and unwieldy fibres into a consolidated, stronger and far more functional form. The Spinning Wheel transformed our forebears' ability to produce threads and yarns on a scale that made the satisfaction of one of our most basic needs practical and accessible to everyone.

What Was Invented?

The Spinning Wheel—a device to produce thread or yarn from a variety of natural or synthetic fibres, for example, cotton or wool—is thought to have originated sometime between 500 CE and 1000 CE in India. Some sources put its origin more specifically at 700 CE. Regardless of the exact date, it appears that it was, unequivocally, an invention that occurred during the Dark Ages period.

The Spinning Wheel of this era is somewhat different in appearance from the more familiar, and more recent, European models. Figure 2 shows a modern version of the traditional Indian design. Very simply, the device aided the slower and more laborious manual process of drawing out and twisting together plant (e.g. flax or cotton) and animal (e.g. sheep's wool) fibres to make yarn. Prior to the Spinning Wheel, the manual process consisted of using a *distaff* to hold the unspun fibre, and a *spindle*, onto which the fibre was drawn out, twisted and coiled. While reasonably effective in producing yarn, this method's chief drawback was its slowness.

The first Spinning Wheel, like its later variants, sought simply to increase the speed of an individual spinner. Instead of turning the spindle with one hand, while feeding raw fibre onto it with the other, the Spinning Wheel simplified the turning of the spindle, also turning it more quickly, and allowed some greater measure of freedom to use both hands to feed the fibre onto the spindle. The old, manual method also usually required the spinner to balance the distaff across one shoulder, so that the entire process was somewhat cumbersome. With the advent of the Spinning Wheel, not only was the spindle taken out of the hands of the spinner, but in some variants, the distaff was also held by the frame of the Spinning Wheel.

With even the most rudimentary, early version of the Spinning Wheel, the spinner could turn the large wheel with a few pushes of one hand, and then use both free hands to feed unspun fibres onto the spindle. This gave greater control of the fibres to the spinner as well as accelerating the process. As the wheel slowed, the spinner could give it a few more pushes, again allowing a period in which both hands were free to feed fibres onto the turning spindle.

Fig. 2 Indian Charkha or Spinning Wheel, Jammu, ca.1875–1940 (*Credit* Wikimedia)

The key purpose of spinning the fibres was, of course, to produce yarn, thread and even cordage (for rope making). Once these materials were produced, weaving, knitting, sewing and other methods could then be used to make a variety of textiles and clothes. These in turn resulted in clothing and other functional materials that satisfied many basic needs, especially in climate conditions that required protection from the elements.

The Spinning Wheel was a clear incremental development of earlier hand-spinning methods and was itself improved and further automated in later eras. Although true mass production of cloth and textiles did not eventuate until the Industrial Revolution brought about the mechanisation of many processes, the Spinning Wheel was an important step on a path of human development.

Why Was the Spinning Wheel Invented?

The Spinning Wheel was invented to improve on an existing process. It seems to be a clear case of incremental innovation. Humankind knew how to spin fibres into thread and yarn, but it was a slow, labour-intensive process. The invention of the Spinning Wheel seems to have been an unequivocal, and successful, attempt to do so better, faster and cheaper. Unlike some of our inventions, there is no secondary or wider problem that directly drove this invention—it was incrementation in its purest sense, and aimed at *how to spin faster*.

How Creative Was the Spinning Wheel?

Relevance and Effectiveness: The Spinning Wheel fully deserves a score of 4 out of 4 for relevance and effectiveness. Prior to the introduction of this device, threads and yarns (the difference here is that yarn is a single strand of interlocked fibres, while a thread consists of two or more yarns twisted together) were produced by a much slower process of hand spinning. The Spinning Wheel supplemented that hand-spinning process by rotating a spindle—the shaft onto which the twisted fibres were wound—using a larger wheel. In this way, through a mechanical process similar to the way a bicycle wheel is turned, the spindle could be rotated more rapidly, enabling the twisting of the fibres to be done far more quickly. While this basic process would be improved further at later points in history—for example, by turning the wheel with a foot pedal—its introduction in about 700 CE must have had a major impact for every culture in which it was introduced. Cloth, whether knitted or woven, was vital in just about every culture, for protection from the elements as much as for more social functions. The faster yarn/thread could be produced, the faster and cheaper cloth could be woven and functional clothing made.

Novelty: Although highly effective when we place ourselves in the shoes of a spinner in this period in India, or later in the Middle Ages when the invention reached Europe, the Spinning Wheel has some weaknesses when we delve into its novelty. Although particularly good at highlighting the weaknesses of previous hand-spinning methods, and offering ideas for further improving previous techniques, the Spinning Wheel, nevertheless, fails to indicate a radically new approach to a strong degree. If we ask "did it break the prevailing paradigm?" then the answer must be "no". It took the core method, twisting and winding raw fibres, e.g. wool, onto a spindle, and added some mechanical aids to this process (but we cannot really even say here that it *automated* this process). There is no doubt, however, that hand spinners would have been delighted by the improvement, and it is easy to imagine that many benefitted, and profited, from this new invention. Therefore, it still possessed a strong element of surprise, and as a result, I give it a score of 3 out of 4 for novelty.

Elegance: The first spinning wheels introduced in India in around 700 CE, and even those that emerged in Europe in the Middle Ages, were not as well executed as the example shown in Fig. 2. Nevertheless, while the *completeness* of the design might be said to be deficient—even to an *ancient* eye it must have looked like there was some room for improvement, for example, in the symmetry of the large wheel and the manner of turning this by hand—the Spinning Wheel must have been sufficiently convincing and graceful that it improved on previous methods, and did not make the process of spinning *harder*. These early spinning wheels must also have been readily understandable to an observer. It is immediately apparent that turning the large wheel causes the spindle to turn rapidly, and that this is beneficial and desirable. At the same time, it must have been a little difficult to use without some practice. Turning the large wheel with one hand left the spinner with only one free hand to feed the raw fibres onto the spindle. Some spinners may have used a second person to assist in turning the wheel, but this would have added to the cost of producing the yarn/thread.

We can see why later developments introduced a treadle to turn the wheel by foot. We also know that a deficiency of these early spinning wheels was that they had difficulty producing the strong and smooth yarns needed for weaving cloth, so again, the execution of the design leaves something to be desired. While a reasonably well-executed solution, it certainly left the door open for many later improvements, and I give it a score of 3 out of 4 for elegance.

Genesis: One of the strengths of the Spinning Wheel is how it stimulated ideas for further improvements. I have already discussed that fact that spinners must have very quickly seen the desirability of turning the large wheel by some means other than their own hand. Not only did this lead to the idea of a foot peddle, or treadle, to carry out that function, but it further stimulates thinking about what parts of the spinning process have to be carried out by the spinner, and what parts could be carried out by someone else, or indeed, through some process of automation. In other words, this early Spinning Wheel was the first simple step on a path of automation and even mass production. It also causes us to think about unrelated problems, such as how *work* (in the mechanical sense of the large wheel turning) is transferred to another object (the spindle). The large wheel and the spindle are essentially two parts of a very simple system of gears, and it's plausible that some of what humankind learned about such mechanical systems was learned through the practical challenge of turning a Spinning Wheel. This knowledge, in turn, was refined and re-applied to other problems entirely unrelated to spinning yarn. For these reasons, the Spinning Wheel scores quite well for genesis and is reasonably strong in changing how we understand a problem or task, and its solution. I give it 3 out of 4 in this category.

Total: The Spinning Wheel, as it first appeared in around 700 CE India, scores a total of 13 out of 16 for creativity. This puts it just into the *very high* range on our scale, and this seems to be a good reflection of its characteristics and impact. It was very effective, even if it had scope for improvement. It was quite novel, quite well executed and shone enough new light on the problem that, while not breaking the spinning paradigm, certainly advanced it considerably towards a point of diminishing returns. While that may seem somewhat negative, in fact it is an important goal for any innovation. Squeeze as much value as possible out of the existing paradigm, and only when the point of diminishing returns has been reached—the maximum value has been extracted—move to a new paradigm, and repeat the improvement process for that.

The Crankshaft (1206)

Automation is going to cause unemployment, and we need to prepare for it – Mark Cuban, American Businessman (1958–)

Our final Dark Ages innovation is an energy-handling system given to us by the Arabic world. Between the eleventh and thirteenth centuries, during the time of the Crusades, many Arabs considered Europeans dull and backward, lacking in scientific knowledge and intelligence. Arabic culture had a far superior understanding of medicine, for example, at a time when European medicine was still mired in superstition.

Many of the technological advances we take for granted nowadays can trace their origins to advances in scientific knowledge that originated outside of Europe during this *dark* period. The crankshaft might seem to be a rather abstract innovation, not obviously tied to the satisfaction of any of Maslow's needs. How is it that an oddly bent, usually metal rod has become so essential to some of modern life's most common activities, and how do so many of these contribute to food production, the provision of water and the generation of heat, not to mention the achievement of some of humankind's most transcendental achievements? Let's find out!

What Was Invented?

Ismail Al-Jazari (1136–1206), who lived in what is modern-day Turkey, was a prolific polymath in the service of the Artuqid dynasty in the eleventh century CE. He is particularly known for *The Book of Knowledge of Ingenious Mechanical Devices*, published in the year of his death. In that book, Al-Jazari set out the details of many mechanical devices and systems, including water pumps, automata and clocks—some invented by him, and others described and improved on—but one of the most important, and enduring, of his inventions is the humble crankshaft. This is a simple mechanical device that converts between reciprocating and rotational motion. We see this most commonly in things like the crankshaft found in petrol and diesel engines. In these, the crankshaft converts the up-and-down (reciprocating) motion of the pistons to rotational motion of the driveshaft, thus enabling the piston engine to turn the wheels of the car. The conversion can also take place in the opposite sense—rotational motion to reciprocating motion—and this is typically seen in certain kinds of compressor.

The crankshaft that Al-Jazari described was, in fact, an example of conversion from continuous rotational motion to linear reciprocating motion (i.e. the opposite of what takes place in an internal combustion engine). Al-Jazari described two particular uses of the crankshaft, both associated with moving or *pumping* water. The first was in a so-called *chain* pump, in which a system of several crankshafts and gears, themselves turned by hydropower, were used to move a continuous chain of buckets through a water source. Each bucket would scoop up some water, lift it and deposit the water in a reservoir at a higher level, before continuing back down to scoop up more water (see Fig. 3). The second was a more sophisticated use of the crankshaft, to create a twin-cylinder, reciprocating piston, suction pump. In this device, we first see the more familiar system of a rotating crankshaft joined to reciprocating pistons by connecting rods. In fact, Al-Jazari's invention was not only noteworthy for the use of the crankshaft but also because it appears to be the first use of suction to create a partial vacuum as the means for drawing water into the pump and is also the

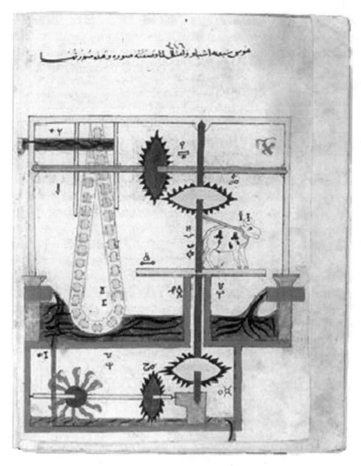

Fig. 3 Al-Jazari water raising device ca. 1205 (*Credit* Wikimedia)

first known example of *double-action* piston mechanism.[3] Al-Jazari's suction pump could lift water over 13 m in height, and while this did not outperform other devices available at the time (such as a waterwheel), it would prove to be far more amenable to the application of external sources of power (such as steam).

The importance of the crankshaft cannot be underestimated. As we will discuss later, the breakthrough of the Industrial Revolution (c1760–1830) was not so much the invention of steam power itself, but the ability to produce rotational motion from steam power. In fact, James Watt's steam engine (developed between 1763 and 1775) initially could not use a crankshaft because that device was subject to patent protection. Once the patent had expired on the crank mechanism, Watt reverted to its use,

---

[3]In a double-action piston, the so-called *working fluid* acts on both sides of the piston. This is common in steam engines but is *not* the process used in an internal combustion engine. In the latter system, fuel and air are only ever present on one side of the piston.

and the potential of steam-powered, rotational motion could be realised. Even after the demise of steam power, rotational motion had remained central to industrialisation, with the crankshaft still an essential component of petrol- and diesel-powered vehicles, among other things. Al-Jazari's invention of the crankshaft some 600 years earlier was therefore very much ahead of its time.

Why Was the Crankshaft Invented?

Ismail Al-Jazari was a practical man, with an intense interest in mechanical systems applied to practical problems. It seems likely, therefore, that he was concerned, or motivated, not by the narrow issue of how to convert reciprocating motion to rotational motion, but with questions such as how to pump water. Nevertheless, his crankshaft must solve a very particular problem in order to solve these bigger issues, and others that he could not have anticipated. Therefore, I suggest that the key problem he tackled was *how to convert motion*. This is an issue of how energy is converted into useful work, and I'll keep it a little non-committal, to allow for the fact that the crankshaft can convert in either direction (reciprocating to rotational, and vice versa).

How Creative Was the Crankshaft?

Relevance and Effectiveness: Like many of our catalogue of inventions, Al-Jazari's crankshaft fully deserves its score of 4 out of 4. It represented a correct and technically sound application of mechanics and physics available at the time, it did what it was supposed to do, in the sense of converting rotational motion to linear, reciprocating motion, and it fitted the constraints of the day. This latter point reflects the fact that the invention was technically feasible, and did not outstrip what was physically possible at the time.

Novelty: While it is clear that Al-Jazari's crankshaft, and indeed, the associated elements such as the connecting rods and pistons used in his pump, clearly demonstrate the shortcomings of other, lesser approaches to the same basic problem, the crankshaft has some minor deficiencies against some of the other criteria associated with novelty. For example, the fact the suction pump could not outperform other existing solutions such as the waterwheel must have cast some doubt on the crankshaft as an improvement. Similarly, because it is mechanically rather complex, it may have been difficult for an observer present at the time to anticipate the benefits that it would bring. In other words, while it is quite strongly radical in its approach, it may have been difficult for people at the time to see other ways that the device might be used. Clearly, people did do this, as history shows us, however, it would be centuries before the crankshaft became the basis of converting reciprocating motion to rotational motion. Nevertheless, the crankshaft represents an enormously influential injection of novelty into the world of energy-handling systems, and still scores a worthy 3.5 out of 4 for novelty.

Elegance: In the crankshaft, and a practical application such as the suction pump, we see an interesting interplay between elegance and novelty (in addition to the more usual impact of elegance on effectiveness). Some of the minor weakness of the crankshaft with respect to novelty may relate to its relative complexity, and in many ways this is reflected in some weakness in elegance. We have evidence suggesting that the crankshaft-enabled suction pump that Al-Jazari invented was skilfully executed, well proportioned and gracefully formed. At the same time, it must have appeared to an observer as complex, somewhat hard to understand and perhaps even somewhat baffling, at least without some explanation, especially set against other water-lifting technology of the day. These may be minor quibbles, but they just shave a little gloss off the crankshaft's elegance, leaving us with a score of 3.5 out of 4.

Genesis: This criterion asks us to consider the extent to which the solution—the crankshaft in this case—changed how the problem was understood. That reminds us that we must be clear about the nature of that problem. Was Al-Jazari concerned with the problem of converting rotational motion to reciprocating motion—quite literally—or was he concerned with the problem of how to lift water. I hypothesise that he was, indeed, specifically concerned with the former, with the latter being a convenient practical application of his invention. In fact, together, these two questions demonstrate a key aspect of genesis—the generalisability of an invention. Once Al-Jazari made the breakthrough embodied in the crankshaft, it must have very quickly led him to further developments of the underlying principle as well as to ideas for solving a range of apparently unrelated problems (like pumping water). Not only that but the basic concept of the crankshaft would have led to the identification of related problems—how to connect a piston to the crankshaft; how to seal a piston in a cylinder—and would have suggested new ways of looking at many existing problems. Across the board, for genesis, the crankshaft must have a highly disruptive influence, changing how the underlying problem of using and manipulating energy was concerned, and leading to an avalanche of new applications. For these reasons, I think the crankshaft deserves the maximum score of 4 out of 4 for genesis.

Total: The crankshaft achieves a score of 15 out of 16, placing it close to the top of our scale of creativity. The only minor deficiencies are found in the novelty of the device—and this is chiefly, perhaps, related to the degree of complexity, and therefore a minor deficiency in elegance. However, in terms of effectiveness, and in how it disrupted energy-handling concepts, it deservedly sits firmly in the *very high* range of creativity.

The Renaissance (1300–1700): The Rebirth of Knowledge

We follow the Dark Ages—perhaps not nearly as dark and unenlightened as the name suggests—with the period known as the Renaissance. This time of *rebirth* began, in particular, with a renewed interest in learning, drawing on classical sources and reforms in the process of education. Another defining characteristic of the Renaissance was the development of linear perspective, which was part of a more general move towards greater realism in painting. A general surge of interest in observation of the world, and the development of inductive reasoning—formulating explanatory theories based on observations of the world—were organised for the first time in a more formal way, and *experimentation* was an innovation in how humankind made sense of the world in which they lived.

Many sources and writers have tried to capture the essence of this period, which is generally regarded as having begun in Italy, with people such as Leonardo da Vinci (1452–1519). The term *Renaissance man* was popularised in the nineteenth century as a glorification of *polymaths*—from the Greek meaning *having learned much*—referring to da Vinci, and other academic and artistic all-rounders such as Michelangelo. Indeed, you may have expected to find something from da Vinci in this chapter. The problem for da Vinci, and a fact widely acknowledged in modern literature, is that very little of what he invented actually made it off the drawing board. As a result, no matter how novel his ideas—and he is often said to have *invented* the idea of the parachute, the military tank and the adding machine, for example—this lack of demonstrated effectiveness and tangible elegance would limit his scores on my scale to 8 out of 16. Rather than insult his memory, or offend his fans, I'll take the diplomatic way out and steer clear of da Vinci's "inventions". A related problem is that da Vinci generally did not publish his own work (and, in fact, is renowned as having been quite secretive) so that his ideas did not directly impact on contemporary thinking. Unfortunately for our purposes, and with our contemporary assessment of creativity, it is remarkably hard to offer any judgements about his work. For us, therefore, he remains simply a brilliant artist, painter of the Mona Lisa and The Last Supper.

© Springer Nature Singapore Pte Ltd. 2019
D. H. Cropley, *Homo Problematis Solvendis—Problem-solving Man*,
https://doi.org/10.1007/978-981-13-3101-5_6

In this time period, we will concentrate our lens on three inventions. The first innovation is a farming implement known as the *scythe*. Unspectacular, but vital to humankind's survival and development, especially in an era in which Europe had been devastated by the Black Death, and where efficient food production suddenly became a hot issue. The second innovation of this era, and a popular one in the minds of many historians of technology, is the movable-type printing press developed by Johannes Gutenberg. Even at the time of its invention, it must have been immediately obvious that this was an important advance. With the benefit of hindsight, we also see the impact that the printing press had, not just on the sharing of information, but on education, democracy and the achievement of humankind's highest potential. Finally, we will round out this time period with an invention that you probably didn't see coming—the *linen condom*! This innovation represents a shift in humankind's understanding of the world. It was during this period that our ancestors began to understand that illness and disease were not punishments meted out by vengeful Gods, but could be understood and treated.

The Scythe (c1300)

Do what you can, with what you have, where you are – Theodore Roosevelt, US President from 1901–1909 (1858–1918)

The scythe may have been invented as far back as 500 BCE, and like many tools, it fulfils the fundamental purpose of an energy-handling system. In essence, the scythe is a device that transforms, or supplements, human energy, allowing our ancestors to perform an action faster, or more strongly, than would otherwise be the case. Thus, a hammer allows us to strike objects more forcefully than we can with our bare hands. A screwdriver allows us to turns objects with greater power than would be possible with our fingers, and a scythe turns the energy of our arms, trunk and legs into an efficient means for cutting things. It does this not to satisfy a need for fulfilment or accomplishment, or to foster intimate relationships, but because our ancestors were hungry, and needed to find efficient ways to produce food, so that this basic need could be satisfied.

What Was Invented?

The basic scythe design consists of a wooden shaft (the *snaith*) of about 1.7 m in length. At one end is attached a curved metal blade typically between 0.6 and 0.9 m in length. Two handles are attached; one roughly halfway along the length of the shaft, and one at the end opposite to the blade (see Fig. 1). The scythe was developed for mowing grass or reaping crops and is operated by swinging the blade from right to left, through the uncut grass or crop that is ahead of the operator. The cut material

Fig. 1 Man using scythe (*Credit* Wikimedia, Gerald Quinn)

is deposited to the mower's left by this action, forming a line of *windrow*. The mower moves forward after each swing of the blade, forming a *swathe* of cut grass/crop, and a continuous line of windrow.

The scythe presents some challenges for our analysis because there is evidence that it existed in a recognisable form possibly as early as 500 BCE in the Classical Period of ancient Rome. There is further evidence of its use during the earlier part of the Dark Ages, with surviving pictures dated at 850 CE showing scythes. What makes it interesting is both the fact that it competed with the sickle—a simpler, single-handed curved blade used for cutting grass and crops—for a long period of the history of farming and agriculture, and because it took so long to replace it, despite obvious advantages.

The sickle was slower and less ergonomic than the scythe. It is estimated that a single farm worker could harvest about 33% more with a scythe each day, and it did not involve bending over uncomfortably in the way that the sickle did. With later improvements, especially the addition of wooden pegs along the length of the shaft (the snaith), grain could be both cut and bundled into sheaves in a single action, further doubling the efficiency of harvesting. It would not be until the invention of horse-drawn, mechanical reaping machines, in the United States during the 1830s, that scythes would be superseded. Even with mechanical reaping machines, scythes continued to be used into the twentieth century in some cultures.

The reason we consider the scythe in this particular time period, and why we focus on the date of around 1300 CE, is twofold. First, it was around this time that scythes began to be used not only for cutting hay but for harvesting crops, and second, because this time period helps to illustrate an important relationship between change and creativity. Prior to the mid-1300s, farming in Europe was a labour-intensive industry. After the Bubonic Plague epidemic that took place in the period 1347–1351—the so-called *Black Death*—European farming faced a crisis. This pandemic killed between 30 and 60% of Europe's total population, with obvious effects on any activity that required large numbers of healthy workers. Efficiency therefore became a key for activities such as farming, if the surviving populations were to be fed, and it was in this period of European history that many agricultural innovations took place. This is the likely driver of the transition of the scythe from mowing hay to harvesting crops and shows the impact of externally imposed changes as a source of innovation.

Why Was the Scythe Invented?

Ever since humankind began cultivating plants as a source of food, there has been a need for innovations to support this process. Cropping, in fact, gives us a very neat illustration of the differences between incremental and radical innovation, linked to the concept of diminishing returns. A farmer grows a crop. This must be harvested in a timely manner if it is to serve as a source of food. Sickles were invented to cut and gather crops such as wheat, so that the grains could be turned into flour. Larger populations required larger crops, and larger crops meant more people to harvest them. One way to harvest a larger crop more quickly is to add more and more people—more sickles—to the process. However, there comes a point where there is no further gain in efficiency as the people get in each other's way—there is a diminishing return. Add into this the fact that there were no extra people (thanks to the Plague), and we see a point where the only pathway forward is not incremental, but radical—a completely new solution, such as replacing sickles with scythes. The scythe was originally invented to solve the problem of how to cut grass. In the 1300s, it was adapted to a new problem, namely, *how to harvest crops efficiently*?

How Creative Was the Scythe?

Relevance and Effectiveness: I mentioned earlier that the scythe is somewhat problematic in our analysis. Some sources put its invention at around 1300, while others offer evidence that it was invented substantially earlier. We can resolve this by suggesting that the conflicting claims may be evidence that it appeared in different places at different times, or was even lost and rediscovered in some cultures. For our purposes, I will assume that it was invented/rediscovered in Europe around 1300 and analyse its creativity from this perspective. On that basis, the scythe certainly epitomises the application of knowledge available at the time. Europe in the early 1300s possessed a reasonably sophisticated knowledge of metalworking—especially iron—which was necessary for the blade of the scythe. The ability also to shape and manipulate wooden tools was similarly advanced, and together, these skills made it possible to create functional tools such as the scythe. This invention also performed reasonably well in its core function of cutting grass/crops. It must lose some points, however, as the scythe nevertheless performs this function less efficiently than must be theoretically possible. It's true that it was an improvement over the sickle—it could cut more grass, more quickly and more comfortably—however, even users at the time would have seen the value of being able to do so *ever more* efficiently and effectively, even if they couldn't find a means to do so. It is also clear that the solution to such an improvement was not simply to make ever-larger scythes. There is a finite limit on the scythe based on the strength and physical attributes of the human operator. For these reasons, I give the scythe a score of 3 out of 4 for relevance and effectiveness.

Novelty: I give the scythe a score of 2.5 out of 4 for novelty. It scores reasonably well for all the criteria of novelty. The scythe highlights many of the weaknesses of other methods of cutting grass and crops. The common alternative, the sickle, was essentially a smaller, one-handed equivalent to the scythe. The swathe that could be cut with a sickle was narrower, and cutting action did not create the windrow as easily, or effectively as the scythe, creating additional work for farmers in collecting the cut material. The scythe ably demonstrated the advantages of a larger, two-handed cutting device, and must also have been considerably more comfortable for the operator to use. However, while the scythe is reasonably effective in showing how the basic principle of a blade could be extended in a somewhat new direction, it is largely incremental in nature, and offers little in the way of redefinition—that is, there are few ideas that emerge for new and different ways to use the scythe. It has one real purpose, and little else. Even as a weapon, it would have been clumsy and difficult to use. While we know that basic foot soldiers of mediaeval armies might often be armed with pitchforks or clubs, we do not seem to have any evidence that they were ever equipped with scythes, although I may be wrong.

Elegance: The scythe scores quite well across the criteria associated with elegance. It is convincing in the sense that even simple, ancient examples of the scythe show a high degree of craftsmanship. It is easy to imagine that a well-made example, properly adjusted to the user, must have been a valued tool. The scythe is also rather pleasing to the eye, with a gently curved shaft lending some added functionality to the device, while the curved blade suggests a design in which some effort has been made to optimise its ability to cut grass and crops, while lifting and depositing the windrow to one side. We also know that there was considerable skill, and practice, involved in using a scythe effectively, suggesting that a well-designed scythe made a difference to the result. For these reasons, I give the mediaeval scythe a score of 3 out of 4 for elegance.

Genesis: Did the scythe break the prevailing paradigm of cutting grass and harvesting crops? Did it disrupt this aspect of farming, leading to the obsolescence of previous methods/tools (such as the sickle)? The answer to these questions is a rather uncertain "maybe"! We can argue that the scythe began a transition from the most basic approach of simply cutting the grass/crop, to one in which the tool also added some value (such as creating the windrow of cut material for easier gathering). The scythe also began a process of attempting to improve the efficiency of farming, which up to this point in history had largely depended on the free availability of human labour. On the other hand, the scythe seems to offer little in the way of a real shake-up. A single person still swung a blade to cut the grass or crop, albeit somewhat more efficiently, but this suggested little that was a fundamental reorientation and offered little in the way of opening up ideas for other unrelated problems. The scythe set a new standard for harvesting, but I find it hard to say that it had a major, disruptive impact. I therefore give it a score of 2.5 out of 4 for genesis.

Total: The scythe achieves a total score of 11 out of 16 for creativity. This is at the lower end of our collection of inventions but remains *high* in absolute terms. We have to guard ourselves against dismissing inventions such as this as uncreative. If you could send me back to the year 1300 in a time machine, I would take a scythe over a sickle in an instant. However, while the scythe is as functional as many of our inventions—and this is always a critical quality for any practical, impactful invention—its weaknesses lie in the two qualities that define newness. The overwhelming sense I have for this aspect of the invention is that the scythe was a modest, positive step forward. Better than what it replaced, but still sitting on the same diminishing returns curve as the sickle, with little more to do to reach that point of diminishing returns. As I stated earlier, there was little more to be done to the scythe to improve it. It could be a little more comfortable and have a slightly longer or sharper blade, but beyond that, the next advance would have to come by stepping onto a completely new curve.

The Printing Press (1450)

The invention of the printing press was one of the most important events in human history
– Ha-Joon Chang, South Korean Economist (1963–)

The Gutenberg printing press addresses higher order needs in Maslow's hierarchy. Although sophisticated writing systems and languages were well and truly established by the fifteenth century, written language was still the domain of educated elites. Few people could read and write, and the means to do so—clay tablets, parchments and so forth—meant that the technology was simply out of the reach of most people. The printing press began a fundamental process of democratising information, by putting reading and writing, and the ideas expressed through written language, into the hands of many more people. Once it was possible for an ordinary person to possess printed documents, it became possible for them both to tap into a pool of knowledge, as well as contribute to it. Although it would take centuries before this process of the democratisation of ideas was complete (if, indeed, it is complete), it can be said to have begun with Gutenberg's printing press.

What Was Invented?

The invention of the movable metal type printing press by Johannes Gutenberg is often presented as one of history's most important inventions, and not without good cause. In fact, a clear case can be made that the printing press had a number of other, more far-reaching effects than simply increasing the accessibility of books.

For example, the printing press started a chain of events that ultimately would lead to a change in the way we understand the ownership of written ideas. For example, many of the classic texts in mathematics were translated into Italian, and published in Italian and Latin, significantly increasing their availability at a time when the Renaissance was taking firm hold. This was also critical in helping to popularise the "new" Hindu-Arabic numbering system that had been introduced to Europe by Fibonacci.[1]

The underpinning concept of the printing press actually predates Johannes Gutenberg's development, in 1450, of a system whereby individual letters were formed from metal alloy, and held in a frame, to then be pressed onto paper. Prior to 1450, printers (the people) relied on craftsmen to carve whole pages into wooden blocks. Not only was this a highly skilled, and therefore slow and expensive technique, but the wooden blocks deteriorated fairly quickly, so that they were unable to print in large quantities. Gutenberg's invention of the so-called *movable metal type* allowed printers to quickly assemble pages of text from pre-formed letters. Not only was this faster and cheaper—the type could be assembled quickly, and the letters were completely reusable—but they lasted far longer than the previous wooden blocks, and thus allowed printers to print many more pages from a single template.

In fact, Gutenberg's contribution did not stop with movable metal type. He also introduced oil-based inks to replace the water-based predecessor and added coloured inks to the toolkit of the printer. This package of incremental innovations quickly demonstrated its utility, not least in the Bible that Gutenberg produced in 1456 (see Fig. 2 for an example).

---

[1] Leonardo of Pisa, Italy (c. 1175–c. 1250), better known as Fibonacci, was one of the Middle Ages' great mathematicians. His 1202 book, *Liber Abaci* (Book of Calculation) popularised the Hindu-Arabic numbering system, using the digits 0–9 and the concept of place value. He demonstrated the utility of this, for example, for bookkeeping. This system began to replace Roman numerals, and readers will appreciate the improvement by trying to do a sum such as 87×9 (i.e. LXXXVII × IX) on paper, using Roman numerals.

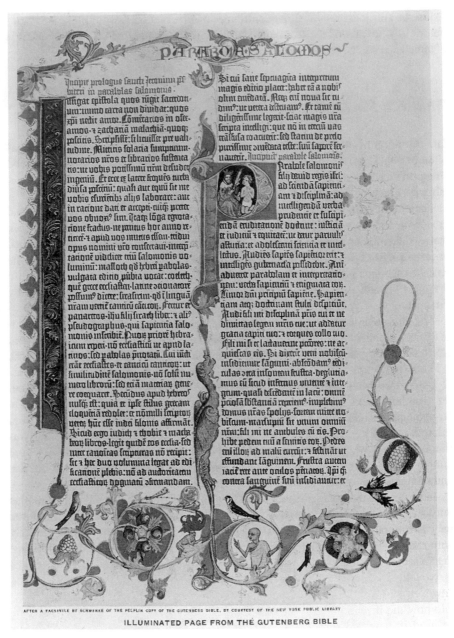

ILLUMINATED PAGE FROM THE GUTENBERG BIBLE

Fig. 2 Proverbs of Solomon, Gutenberg Bible, fifteenth century (*Credit* Wellcome Collection)

In many respects, therefore, Gutenberg's inventive efforts not only improved printing but also greatly increased the value and potential of *writing* (see the discussion of Cuneiform writing in the Prehistory era), *paper* (see the discussion of this technology in the chapter on the Dark Ages), and *education* directly, and many of our other innovations at least indirectly.

Why Was the Printing Press Invented?

Gutenberg's movable metal type printing press was not the first solution to the problem of sharing written information. For centuries, scribes wrote out by hand documents that needed to be copied and shared. Wooden block presses improved on this process in some respects but remained slow and labour-intensive in many respects. It was Gutenberg's incremental improvements that resulted in a quantum leap for printing. Suddenly, documents could be produced far more quickly, and in far greater volume, than had been possible, and this uncovered a thirst for written information. If the preceding technologies solved the basic problem—how to share written information—then Gutenberg's printing press added "quickly, cheaply, efficiently" and other qualifiers to this statement. Gutenberg made printing *practical*.

How Creative Was the Printing Press?

Relevance and Effectiveness: Gutenberg's innovation must have immediately made its impact felt. The improvements in speed and quality, and related developments in the cost of printing, were highly relevant—or task appropriate—and undeniably effective as a solution to the problem of distributing and democratising language and knowledge. Gutenberg directly addressed two weaknesses of printing technology—the production of the templates and the quality of the finished product—in what must have been seen as spectacular fashion. Although incremental in nature, these were big increments, and must have had the feel, and the effect, more commonly felt with radical, paradigm-changing innovations. It is impossible to give anything less than a score of 4 out of 4 for this criterion.

Novelty: 3 out of 4 may seem like a low score for the movable-type printing press, however, part of novelty is the extent to which the invention opens up new insights into how the invention—the printing press, in this case—is used, and how it might be used in new and surprising ways. I argue that Gutenberg's ideas, as effective as they were, represent a step improvement to the state of the art, and not really a fundamental change to the state of the art. They were surprising and new in a more inward-looking sense. We might even argue that they took printing almost to the point of diminishing returns—where little further gains could be made, without a fundamental paradigm shift. That's a good outcome, and there was novelty in good measure, but it probably didn't step outside the existing paradigm. If there is a God of printing, may it forgive my blasphemy!

Elegance: I resolved, in my introduction, to take the point of view of a person present at the time these innovations were first introduced. Unfortunately, sometimes it's hard to ignore a peek into the future. In the case of the movable-type printing press, and elegance, my gut-feel is that even Gutenberg must have felt that his system had some room for improvement. To the fifteenth-century eye, it must have been an impressive piece of machinery, but problem-solvers are nothing if not perfectionists, and I have a sense that printers of the day must have seen at least little things in Gutenberg's device that could be executed a little better. Indeed, Gutenberg himself continually tinkered with his invention(s) suggesting that they weren't quite complete, and fully worked out. A good solution usually looks like a good solution, and the movable-type printing press was pretty good, but not perfect. Therefore, I am giving it a 3 out of 4, as much to encourage Johannes, as anything else. Our modern eye also knows what's in store, and there were some big improvements still to come for the whole process of printing.

Genesis: A score of 2.5 out of 4 for genesis feels mean, but if these ratings are to have any value, I have to be as impartial and objective as I can be. In this case, it's easier to explain what is slightly deficient about this invention, rather than praise what is good. One weakness, related the highly incremental nature of the innovation, is a relative lack of offering ideas for solving unrelated problems. Remember that highly creative solutions often solve problems we weren't even aware of. The printing press solved the obvious problem, but I find less connection to other problems. Of course, we can argue that this is not the case, and that probably depends on how big you are willing to extend the problem space—where do you draw the boundary? The other slight weakness in genesis is the extent to which Gutenberg's innovation suggests a new basis for further improvements. My feeling is, as I have already indicated, that in this invention we reach a pinnacle—not a bad thing—but genesis needs us to peek over that pinnacle, and see what's on the other side.

Total: It is always tempting to second guess myself with these ratings. However, I have tried very hard to maintain objectivity. One way I do this is that I jump around with the assessments—I have not done them in chronological order, and I have not used them to select the inventions included in this book. A total of 12.5 out of 16 is in that dead zone between high and very high. Objectively, there is nothing wrong with that, but it seems a tough call on an invention like the movable-type printing press. In the assessment of genesis, I hinted at a reason for this discomfort. If we are willing to look at a system in the biggest possible sense, then the printing press seems very influential—the effect on society, education, knowledge and so forth is almost impossible to estimate. So, perhaps we also need a score for each innovation that we consider, which captures the wider influence of that innovation. This is typically only revealed after the fact. If we were to do that for Gutenberg's invention, then I feel it must fare very well.

The Condom (c1560)

Expired condoms are like nuclear waste: there's nothing sensible you can do with it – Andrew Smith, American Author (1959–)

In the last of our inventions of the Renaissance period, we turn to medicine as our field of interest. Although it would be some time before medical science progressed to a more sophisticated theory of the underlying causes of disease, the explosion of new knowledge seen during the Renaissance began to open the door to a more scientific approach to medicine. For centuries, humankind had been at the mercy of diseases. However, as population densities increased, and people lived in closer quarters, and with the greater mobility made possible by ships, many diseases that were previously relatively inconsequential began to make their presence felt. Sexually transmitted infections have often had a devastating effect on societies that previously had not encountered them, although Europeans were usually the transmitters, not the receivers. However, syphilis was an exception, and the invention of the linen condom—a material-handling system—directly addressed the basic, physiological need for good health.

What Was Invented?

Chemical-impregnated linen sheaths, designed to cover the glans (readers can make good use of the Web in this chapter!), and held on the user with ribbons, were first described by the Italian physician Gabriele Fallopio (see Fig. 3). Although Fallopio focused his anatomical and medical studies mainly on the head (the cranium, not the glans!), he is perhaps most familiar to us as the person responsible for naming the fallopian tube that connects the ovary to the uterus in the female reproductive system.

Fallopio's interest in the condom resulted from his study of the sexually transmitted infection syphilis. In his treatise on syphilis—*De Morbo Gallico* (On the French Disease)—which Fallopio published in 1564, he described the linen condom, and also discussed a trial of the device involving 1100 men, none of whom, he claims, contracted the disease.

Although stimulated by the appearance of syphilis in Italy—in other words, the driver for the condom was disease prevention—by the early 1600s, it appears that their use was widespread also as a means of birth control.

Why Was the Linen Condom Invented?

The first documented outbreak of syphilis in Europe occurred in the 1490s. This appears to have been the stimulus behind the development of condoms of a type

Gabriele Falloppio (1523–62).

Fig. 3 Portrait of Gabriello Fallopio (1523–1562) (*Credit* Wellcome Collection)

that is now universally familiar. The direct problem that drove Fallopio's inventive powers was quite simply, how to prevent syphilis.

How Creative Was the Linen Condom?

Relevance and Effectiveness: The linen condoms described by Fallopio, designed for the purpose of protecting against syphilis, seem to have been extremely effective, if Fallopio's trial data are accurate. While we don't have any clear or objective data on the protective benefits of the linen condom against other STI's, and notwithstanding the prevalence of such infections in the sixteenth and seventeenth centuries, it seems reasonable to assume that the linen condom tackled the problem well. Not only did it do what it was supposed to do—stop syphilis—but it accurately reflected the knowledge of the cause (intimate sexual contact), and it appears to have satisfied the obvious constraint, namely, permitting sexual intercourse to take place whilst wearing it. While it may look rather ungainly to a modern eye, I give the linen condom a score of 4 out of 4 for effectiveness.

Novelty: The linen condom scores particularly well for some aspects of novelty. For example, it draws full attention to weaknesses in other, existing methods of the prevention of sexually transmitted infections like syphilis. Indeed, apart from abstinence, there was no effective alternative method of prevention. The condom therefore decisively shows how to improve existing methods of protection, and quickly established the likely effects of its use. At the same time, the concept of covering the penis in some way was not unfamiliar—this had been done previously in other cultures, but for the primary purpose of birth control—so that the condom was not entirely new in concept. It can be said, however, that the condom described by Fallopio did offer new and different ways of using the underpinning sheath concept, and offered a new material (linen). Previous examples of the protective sheaths found in other cultures, and designed for birth control, used materials such as oiled silk paper or tortoiseshell. For these reasons the linen condom gets an impressive, if not quite perfect, 3.5 out of 4 for novelty.

Elegance: Placing ourselves, as is our practice, in the shoes of someone present at the time, the linen condom scores quite well in terms of how well executed a solution it is. We have to keep in mind that Italy (Fallopio's home), and specifically the city of Naples, had experienced a major outbreak of syphilis in 1494/95 thanks to a French invasion. While there is scientific debate as to the origin of syphilis—some claim it was brought to Europe by sailors returning from Columbus's exploration of the New World (in 1492/93), while others say it originated in Europe—its lethality in Fallopio's day was considerably greater than it is today. As far as the elegance of the linen condom is concerned, people of the day must have seen it as a reasonably skilfully executed and complete solution to the problem of syphilis infection. At the same time, it may not have been the most romantic, and comfortable (for either party) solution; however, its primary purpose was protection and not pleasure! Although we avoid drawing on our knowledge of later developments in these analyses, the fact is, the linen condom would compare favourably to later alternatives. Sheep intestines, while thinner and presumably more comfortable, were prone to breaking and contain tiny pores through which some bacteria can pass, while early rubber condoms (1855) produced in the years after the invention of the vulcanisation process by Charles Goodyear, were as thick as bicycle inner tubes! Therefore, while not perfectly elegant in terms of aesthetics and execution, the linen condom scores a healthy 3 out of 4.

Genesis: As has been the case for many of our inventions, we once again find ourselves faced with an innovation that is more incremental that radical. The underlying concept of a thin protective cover on the penis preventing direct contact and the exchange of biological material between male and female sex organs was not new. The particular application—for the prevention of sexually transmitted disease—*was* new, and the linen condom also serves as a foundation for further work on this same problem. Having established a basic *barrier* concept, the obvious question is, what other materials (that may be better suited to the purpose) can be used to make a condom? This barrier concept might even be said to offer ideas for other unrelated problems. The linen condom also, for example, reminds us of the face masks routinely used by medical professionals. These create a simple cloth barrier to limit the transmission of bacteria and viruses that might be expelled by the wearer, or inhibit the inhalation of the same. As a reasonably strong example of an innovation that changes perspectives and gives rise to new ways of looking at a problem, I give the linen condom a score of 3 out of 4 for genesis.

Total: The linen condom first described by Gabriele Fallopio achieves a creativity score in the *very high* range with 13.5 from the maximum of 16. Although scoring the maximum for effectiveness, the linen condom loses a small amount against the criteria novelty and genesis, largely stemming from the fact that the concept already existed, even though it was realised using different materials and for a different purpose. Aside from that, the linen condom also gives up a point against elegance for some weakness in execution. Nevertheless, the score of 13.5 compares very favourably, if a little unexpectedly, with the other innovations in our catalogue, and is an example of how creativity and innovation can often be found in the most unexpected of places!

The Age of Exploration (1490–1700): New Worlds and New Problems

The Age of Exploration—sometimes also referred to as the Age of Discovery—was a period during which the Western world began to look outwards. The rebirth of knowledge and humankind's interest in its own place in the world marked by the Renaissance set the stage for the application of much of what had been learned since 1300. This period, covering the time from the end of the fifteenth century to the close of the seventeenth century (roughly 1490s–1700), saw many great geographical discoveries. Christopher Columbus sailed west, ultimately opening up both South and North America to settlement by Europeans (not without many problematic consequences). Magellan led the first expedition to circumnavigate the Earth in the early 1500s, and the great Southern continent was first sighted by explorers such as Abel Tasman.

Not surprisingly, this surge in exploration and discovery proceeded in parallel with a surge in invention. Exploration, for example, required accurate navigation, and accurate navigation needed better methods for telling the time and for calculating the positions of the planets and stars. Therefore, our first invention in this era is the *slide rule*. A seemingly rudimentary information processing system, and yet one which allowed our forebears to make much more accurate, and rapid, arithmetical calculations that aided not only navigation, but science, engineering and a range of other problem-solving activities. The second innovation we consider is the *pendulum clock*. Although this would turn out to be unsatisfactory as a solution for accurate timekeeping at sea, and therefore not much help for navigation, the pendulum clock became the gold standard for accurate timekeeping, in all other applications, for over 270 years. We end this era with a third innovation and one which was the herald of many of the advances that would follow it—the *steam pump*. This early *engine* first demonstrated the power of steam as a controllable means of transforming almost all of our ancestors' activities, from farming, through manufacturing, transport and many other facets of society.

© Springer Nature Singapore Pte Ltd. 2019
D. H. Cropley, *Homo Problematis Solvendis—Problem-solving Man*,
https://doi.org/10.1007/978-981-13-3101-5_7

The Slide Rule (1622)

Consequently, like the Blackbirds it powered, the J58 [Pratt and Whitney Jet Engine] was essentially designed by slide rule – Peter W. Merlin, Aerospace Historian (1964–)

We now turn our attention to inventions that sit at the intersection of two overlapping eras: the Renaissance and the Age of Exploration. The first of these inventions is the slide rule, and while it is not the first device designed to assist humans in performing arithmetic calculations (think of the abacus, for example), it makes a strong claim as the first device to automate, in some fashion, the process of making more complex mathematical calculations. The slide rule is, very clearly, an information handling system. As a very general tool, it contributes to the satisfaction of a range of human needs. Unlike a condom, or a scythe, both of which have a much more specific purpose, the slide rule functions as a tool for helping humans to solve all sorts of other problems, and we should expect to see this reflected in its creativity score.

What Was Invented?

William Oughtred (1574–1660) was an English mathematician and Anglican priest who is credited as the inventor of the first slide rule. Some readers may be familiar with a slide rule, and indeed, may even have used one in the days before electronic pocket calculators became widely available (speaking from experience, this was roughly the late 70s/early 80s—I just caught the tail end of slide rules). As an aside, it was the development of a pocket calculator that provided the catalyst for the development of the first integrated circuit—a silicon chip combining a number of different functions on the single chip, in 1960. The slide rule was one of the first devices designed to assist in performing mathematical calculations. I say one of the first because it was preceded by a device called Napier's Bones (in 1617), which assisted in multiplication and division, and of course, by the abacus, which was invented in ancient times to assist with tasks such as addition and multiplication.

The slide is noteworthy for two reasons. First, unlike Napier's Bones, it has survived into modern times and is still readily available. While it is true that the abacus is also readily available, the second advantage of the slide rule is that it allows for a wider range of otherwise difficult calculations—in addition to multiplication and division, it can be used to calculate exponents, logarithms, roots and can also assist in trigonometric calculations.

Oughtred's first slide rule made use of the invention of Natural Logarithms by John Napier (first published by him in 1614), their development into Decimal Logarithms (a task delegated by Napier to Henry Briggs, and achieved in 1616/17) and the subsequent invention of Gunter's Scale, by Edmund Gunter. Logarithms, very simply, represent a form of mathematical notation that greatly improves the accuracy of complex calculations such as multiplication and division, and the calculation

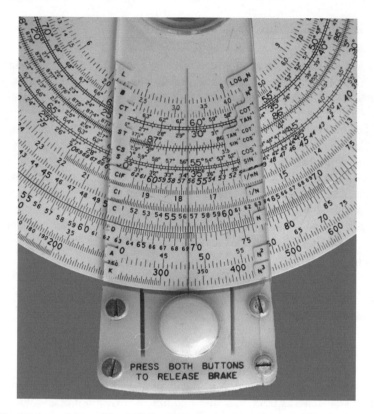

Fig. 1 Partial photo of front side of the Boykin Rotarule Model 510, ca. 1969 (*Credit* Wikimedia, Richard Straka)

of trigonometric functions that are vital to accurate navigation. However, without some form of automation, logarithms are slow and tedious to use. Gunter's scale was the first attempt to automate calculations based on logarithms, by printing a logarithmic scale on a two-foot-long wooden ruler. However, Gunter's scale was still rather unwieldy, and Oughtred gets the credit for making logarithms easy to use by improving on the Gunter Scale.

Oughtred's device was originally circular in form (similar to the modern equivalent depicted in Fig. 1), and consisted of two concentric rings with scales printed on them, and which could be moved relative to each other. If this all sounds horrendously complicated, then take my word for it. This was a big deal! It was also, in effect, the first *analog* (as opposed to *digital*) computer. Until the early-mid 1970s, every self-respecting engineer owned a slide rule. The contribution, of course, was to all of the things that could be done with the aid of more precise mathematical calculations. In this way, Oughtred's slide rule satisfied a necessary precondition for many of the technological advances that would follow.

Why Was the Slide Rule Invented?

Once again, we have a situation where there is a very narrow, specific answer to the question of the problem that was solved. In this case, it seems hard to go past the question *how to simplify mathematical calculations*. However, there seems little doubt that mathematicians like Oughtred were also acutely aware of the driving forces behind that problem. The Age of Exploration tells us something about these driving forces. Countries like Britain, at this time, were expanding their horizons, venturing further and further afield as they explored the globe. One of the first problems that this creates is a need for accurate navigation, and that creates two problems. First, how to measure time accurately, and second how to carry out the relatively complex calculations needed to translate the positions of the Sun and stars into a position on a map. The first would eventually be solved by an English clock-maker—John Harrison—in the 1760s, while the second would be greatly aided by devices such as Oughtred's slide rule.

How Creative Was the Slide Rule?

Relevance and Effectiveness: William Oughtred's circular slide rule scores the maximum of 4 for relevance and effectiveness. It is clear that the device correctly and accurately embodied the available knowledge of mathematics, and specifically logarithms, available at the time. In addition to this, it correctly utilises knowledge of techniques of wood and metalworking to create the actual device itself. This first true slide rule also did what was intended. It allowed a user to conduct mathematical calculations using logarithms in a faster and more accurate manner than was possible through manual, or unaided, calculations. Last, the device conformed to what we can imagine were practical constraints—for example, it had to be reasonably simple to manufacture, and accessible to potential end-users.

Novelty: The circular slide rule invented by Oughtred presents something of a mixed bag when judged against the criterion of novelty. It is particularly strong against the criterion of *prescription*: it showed, in highly concrete terms, how to improve on its predecessor (Gunter's Scale). Conversely, it is hard to credit it with a strong element of *reinitiation*; it is more of an incremental improvement over its predecessor than a radically new approach. Against other factors in the category—e.g. in how it highlights the shortcoming of other artefacts, or in how it may or may not offer a fundamentally new perspective on the problem of automating and supplementing mathematical calculations—it is a fairly strong contributor. Across the spectrum, therefore, I have given Oughtred's slide rule a score of 3 out of 4 for novelty.

Elegance: Oughtred's slide rule scores well for elegance. Although it is difficult to track down images of an original, examples of other similar artefacts made in roughly the same era (for example, the *Circles of Proportion*, made by Oughtred himself in 1648) show beautifully crafted devices made from brass, and carefully engraved. Taking examples such as these, we can confidently speculate that the circular slide rule was well-executed, complete and nicely formed. As was the case for many artefacts that served a useful purpose in this era—e.g. clocks—considerable craftsmanship was often lavished on these devices. Not only for aesthetic purposes, but also because devices such as these were often used, for example, in ships at sea, and had to be robust and long-lasting. The circular slide rule also reinforces our relationship between elegance and effectiveness. It is not hard to imagine that the better made the slide rule, the more accurate it was, and the longer it lasted—in a very real sense, one that was carelessly made, for example, might have inaccurate markings that would lead to inaccurate calculations. For all these reasons, I give the slide rule 4 out of 4 for elegance.

Genesis: As was the case with novelty, I find a rather mixed picture for genesis and the slide rule. Some of the indicators of this criterion are quite strongly present in Oughtred's device, while others are somewhat weaker. *Seminality*, for example, seems to be strongly present in this invention. Not only does the circular slide rule draw attention to what was inadequate in Gunter's Scale, but it also reveals some previously unnoticed problems. This could be as simple as the layout of the device. A circular slide rule is faster to use for multiple calculations, compared to a long, and linear, device. The slide rule also helps us to see new ways of proceeding beyond the slide rule. For example, what other mathematical functions can also be incorporated in this device, or can the process of augmenting or automating the calculations be automated even further? Conversely, the slide rule seems to be somewhat more limited in its *transferability*. Its strength was fundamentally the calculation of logarithms, and it is hard to see what other unrelated problems and solutions might be hiding in this device. I don't mean the solution of mathematical problems, but the solution of entirely different things. In sum, therefore, I give the circular slide rule invented by William Oughtred a score of 3 out of 4 for genesis. Like many of our inventions, it represents a significant improvement over existing artefacts, and even pushes at the boundaries of the prevailing paradigm, but does not really break out of the existing concept into something wholly new.

Total: The score of 14 out of 16 puts the slide rule firmly into the *very high* range for creativity. It is highly effective and highly elegant—two qualities that together make for an extremely *functional* solution. We find some slight weaknesses in novelty and genesis that together illustrate a tendency towards incremental improvements, rather than paradigm-breaking, or highly disrupting, shifts. It is true that Oughtred's slide rule did more than just improve on Gunter's Scale in the sense of making it easier to use, or more compact. It did shift somewhat to a new concept by adding a second moving scale, but I feel that this is still more of a *big* incremental improvement, rather than a wholly new concept. The general conclusion, however, is that this is one of our more innovative artefacts.

The Pendulum Clock (1656)

It's not that we have little time, but more that we waste a good deal of it – Seneca, Roman Philosopher (4 CE–65 CE)

Although it could be argued that the Pendulum Clock is an energy-handling system—transforming gravitational potential energy stored in the clock's weights into kinetic energy in the movements of the clock's various cogs, gears and pointers—I think we would all agree that it is as an information handling system that we see its true value. In a similar fashion, we don't think of a laptop as a device for converting chemical energy into electrical energy! It is also telling that two of the three inventions in this era are information handling systems. Not only that, but both would be central to the development of exploration, navigation, trade and a range of other activities that preceded the First Industrial Revolution. In terms of the satisfaction of needs, the pendulum clock is rather like the slide rule. Its value lay not really in telling the time, rather, it was in what accurate timekeeping made possible across the spectrum of human needs.

What Was Invented?

It was the Dutch polymath[1] Christiaan Huygens (1629–1695)—famous as a physicist, mathematician, and astronomer—who added the pendulum clock to his catalogue of inventions, in 1656. Prior to this device, clocks suffered from relatively poor accuracy, with the best still losing or gaining up to 15 min per day. The pendulum clock, however, improved this accuracy roughly 60 fold, from 15 min (or 900 s) per day to only 15 s per day! The implications of this improved timekeeping are hard to underestimate. Global trade and exploration, so heavily dependent on maritime navigation, ultimately depended on accurate timekeeping. While pendulum clocks did not entirely solve this problem (pendulum clocks lose accuracy when placed on unstable ships), they did contribute to a general improvement in navigation, especially on shorter, coastal voyages. As Europe moved into the Industrial Revolution in the century following the invention of the pendulum clock, accurate timekeeping was also critical for the organisation of work and the growth of modern, industrialised societies. Indeed, although there were many improvements over the subsequent 270 years, it was not until the invention of the quartz clock (drawing its timing accuracy from the precise oscillation of an electrical circuit regulated by a quartz crystal), that the pendulum clock began to be replaced.

Huygens's design was actually derived from a key discovery by the Italian polymath Galileo Galilei (1564–1642), who discovered the principle of isochronism in

---

[1]The term Polymath refers to a person with a very wide and deep range of knowledge. Famous polymaths include *Renaissance Man* Leonardo da Vinci, Galileo Galilei, and twentieth century philosopher Bertrand Russell.

pendulums—in simple terms, this means that the period of swing of a pendulum is approximately the same, regardless of the size of the swing. The basic design of the pendulum clock uses a gear wheel, which rotates, thanks to the force supplied by a weight connected to it. However, the gear wheel is prevented from turning in an uncontrolled manner by a small lever called an escapement. This lever rocks from side to side, and with each rock, allows the gear wheel to turn a small amount. The pendulum is the key to this; it controls the rocking of the escapement. So, the pendulum swings back and forth in a very reliable and consistent time pattern (thanks to isochronism), and, as it does so, it controls the rocking of the escapement (also in the same precise manner). This, in turn, controls the turning of the gear wheel in a precise manner, and accurate timekeeping is made possible. To avoid the pendulum slowing down and stopping, the escapement also gives the pendulum a small nudge with each rock (the energy for this coming from the weight), so that as long as the clock is wound (i.e. the weight is able to fall under the influence of gravity) the clock will keep ticking accurately.

Of course, there were improvements to be made to Huygens's design. One notable change was that in the first designs, the pendulum swung between 80 and 100°. Huygens himself discovered that this introduced some inaccuracies (basically, isochronism doesn't hold for large swings), and new designs, with new types of escapement mechanism soon addressed this, and a more normal pendulum swing of 4–6° was introduced. With further refinements, the pendulum clock would remain the gold standard for timekeeping until the introduction of quartz clocks in 1927, more than 270 years after their invention!

Why Was the Pendulum Clock Invented?

The pendulum clock tackles a very straightforward problem—how to keep time—and we can add in the "how well" criterion of "accurately" if we want to refine that problem statement. I have already made the case that this was vital to such national endeavours as exploration, navigation and trade, but we know from our own daily routines, that accurate time measurement governs just about everything that we do. What is particularly interesting is what people did before 1656, and the wide availability of accurate timekeeping?

How Creative Was the Pendulum Clock?

Relevance and Effectiveness: The pendulum clock given to us by Huygens was extremely important, and as I have already noted, brought about an immediate 60-fold improvement in timekeeping accuracy. While, in general terms, we must, therefore, say that it was highly fit for purpose, I have deducted a small component of the score to reflect the fact that it still did not function well enough to solve one of the most intractable problems of the seventeenth and eighteenth centuries, and that was accurate timekeeping at sea. Pendulum clocks depend, for their accuracy, on a highly stable mounting. The problem with ships is that as they pitch and roll, they interfere with the accurate and regular motion of the pendulum, and therefore the gains seen on land did not flow through to the maritime environment. This basic problem, accurate timekeeping at sea, which is crucial for accurate navigation, was so important that many sovereigns and governments offered substantial rewards for the invention of an accurate maritime clock. The most famous of these prizes was that offered by the British Government's Board of Longitude in 1714. This offered three rewards for different levels of accuracy, ranging from a prize of £10,000 (about US$1.7 m today) up to a prize of £20,000 (about US$3.4 m). Huygens's pendulum clock, even with substantial improvements over the next 100 years, could not claim any of these prizes. For this reason, it gets a not quite perfect score of 3.5 out of 4 for relevance and effectiveness.

Novelty: Huygens's invention, through its huge gains in accuracy, drew clear attention to the shortcomings of other types of clocks, and showed with decisive clarity, how other clocks could be improved. The pendulum clock made substantial incremental improvements to existing clock technology, extending an emerging knowledge of the properties of pendulums to a practical application. The approach used as the core method of setting the timing of the clock—the pendulum—was substantially new and different. The only real weakness in this category is a slight deficiency in the extent of the change in perspective. Clocks were still mechanical devices, and the novelty was to a sub-element of the clock, rather than to the system as a whole. Nevertheless, within the limits of rounding errors, it is hard to go past a score of 4 out of 4 for novelty, for the pendulum clock.

Elegance: One of the standout features of Huygens's pendulum clock was the quality with which it was constructed. In fact, Huygens himself did not construct it, but instead left that task to an expert clock-maker—Salomon Coster—and his skill is reflected in the elegance of the finished product. Like all elegant inventions, Huygens's design also demonstrated the relationship between a well-made device, and one that also functions effectively. We can see from the drawings that still exist (for example, Fig. 2) the care and attention to detail present in the designs. In fact, because of the skill required to make clocks in this period of history, they were regarded as status symbols—only the wealthy could afford them—and this resulted in great care and some artistry being applied in their design. This means that the pendulum clock was not only skilfully executed, but was also pleasing to the eye, and was packaged in a complete manner in what could be described as an early form of user-friendliness. For these reasons it is easy to give Huygens's pendulum clock the maximum score of 4 for elegance.

Genesis: I have given the pendulum clock a score of 3 out of 4 for genesis. If a truly *genetic* innovation is one that breaks with a prevailing paradigm, and changes how the problem or task is understood, then the pendulum clock is a strong, but not perfect, contender. The pendulum clock concept presents a reasonably novel basis for further work, showing that there were new and better ways to implement the core timing mechanism. On the other hand, this mechanism remains anchored in

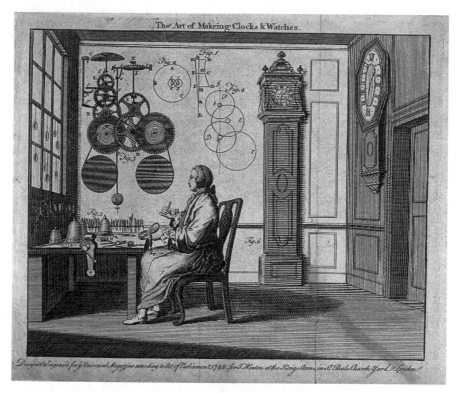

Fig. 2 Clocks: a watch-maker seated at his workbench (*Credit* Wellcome Collection)

mechanics, as opposed to shifting to the use of an electronic mechanism, so that the paradigm shift is partial, rather than wholesale. The pendulum clock also contains a reasonably strong element of *germinality*, demonstrating, for example, that gravity might have exploitable qualities for other purposes. The pendulum clock also scores strongly for the fact that it draws attention to previously unnoticed problems. This is especially the case, for example, with the behaviour of pendulum clocks on ships. Until the introduction of Huygens's innovation, no thought had been given to the interaction of a pendulum's motion with the motion of the structure holding it. This created a new problem that had to be solved by some other means, and provides some stimulus for further innovation. Taking these considerations into account, a score of 3 seems a fair reflection of the pendulum clock's genesis.

Total: The pendulum clock, first introduced by Christiaan Huygens in 1656, achieves an overall score of 14.5 out of 16 for creativity. This places it well into the *very high* range, and indeed, it appears to be one of the strongest contenders in our study for a perfect score. Its only real deficiencies are its relative inaccuracy, and the fact that it did not completely break the prevailing timekeeping paradigm. If it were not for the fact that pendulum clocks could not tackle the Longitude Prize, even after one hundred years of further development, I would be tempted to give it a score of 4 for effectiveness. There can be no doubt that its vastly improved accuracy, 60 times better than any competing clock, must have been almost miraculous in its day. Normally, this would seem to be everything that we need to give it a perfect score for its function, but that Longitude Prize just nags at the back of my mind. This was so important to trade and development at the time that there would have been great hope placed in the pendulum clock as a solution to that problem. Perhaps related to this is the slight weakness in genesis. If the design had *completely* broken the existing paradigm—essentially, that fact that clocks were mechanical devices—by introducing a non-mechanical element to the design, or if it had leapt further forward in terms of the quality of the mechanical works, then this level of genesis would probably have solved the key problem on accurate timekeeping for navigational purposes. Close, but not quite close enough, for Mr Huygens!

The Steam Pump (1698)

I sell here, Sir, what all the world desires to have: Power – James Watt, Scottish Inventor (1736–1819)

Thomas Savery (1650–1715) is considerably less well-known than his compatriots Thomas Newcomen (1664–1729) and James Watt (1736–1819), however, it was Savery's invention of the *steam pump* that laid the foundation of the coming revolution in energy-handling systems. Although Savery's immediate objective was quite specific—he was trying to tackle the problem of pumping water out of mines—we know that he had broader interests, all of which revolved around a growing body of knowledge in science and engineering that was characterised by the Renaissance, and the period that followed. Although I have included the steam pump as part of the Age of Exploration, it is as much a product of the Renaissance.

In terms of Maslow's hierarchy of needs, Savery's steam pump also seems to represent a transition from innovations that were, perhaps, more exclusively focused on lower level needs, to innovations that not only satisfied lower level needs but also opened up possibilities at higher levels. Perhaps that is a fundamental outcome of the Renaissance? Humankind's needs were developing, supported by a rebirth of knowledge, from the more basic to the more sophisticated. From this era onwards, we seem to see more and more inventions that address a wide range of needs. As our forebears got better and better at satisfying basic needs, they gave themselves more time to turn attention to higher order needs.

What Was Invented?

Many readers will be familiar with the first steam engine producing continuous rotary motion, invented by James Watt in 1781. Although Watt's engine may have had a more direct and lasting impact on the Industrial Revolution (which we will consider in the next chapter), it was Savery's invention of a steam pump, sometimes referred as an 'engine', that was the first commercially applied, steam-powered device, nearly 100 years earlier.

Savery was a military engineer who dabbled in experiments in mechanics. Military engineering in this era was primarily concerned with transport, logistics, fortifications, siege engines, and the like. Indeed, Savery patented an invention, in 1696, for manually propelling ships, using paddle wheels and a capstan, but this was rejected by the Royal Navy. It is apparent that he had more than a passing interest in ways of replacing unreliable sources of power, such as the wind, because one of his aims for the first steam pump was to replace wind and water power used in mills, to allow for continuous operation.

Savery described his invention as *The Miner's Friend; or An Engine to Raise Water by Fire*, in his 1702 book of the same name (see Fig. 3). The device, first patented by him in 1698, and demonstrated to the Royal Society the following year, heated water in a boiler to create steam. This steam was then allowed to enter a metal tank by means of a tap. At the bottom of this tank was a pipe connected to the water that was to be pumped. The steam, therefore, filled this tank and blew into the reservoir of water. Once the tank was filled with steam, the tap to the boiler was shut, and the tank was cooled externally. As the steam in this tank cooled, a partial vacuum was created, causing water from the reservoir to be sucked into the tank through the pipe at the bottom of the tank. Readers can experience this basic principle by pouring hot water into a glass bottle, sealing it with an air-tight stopper, and then allowing the water to cool. When you open the bottle, there is a short in-rush of air, showing that the pressure in the bottle has decreased as the hot water/air cools—some suction has been created. Once water had been drawn up into the tank in Savery's design, a tap on the bottom pipe was shut, and steam was then readmitted to the tank, this time forcing the water out of the tank through a pipe on top of the tank. In this way, water could be raised from a lower reservoir, through the tank, to an upper reservoir.

Savery's steam pump had a number of limitations. It was inefficient and could only raise water from the lower reservoir up to a maximum of about 9 m. Although it could, in theory, raise the water much higher when pushing it through the upper pipe, the steam pressure required to do this was difficult to achieve with the limited quality of the seals, pipes and tanks available at the time. However, the fact remains that Savery's innovation was used for pumping water out of mines, and could also be applied to assist in supplying towns with water. It had no moving parts, aside from the taps, and found real, practical applications, despite its flaws. In this sense, Thomas Savery was one of the true catalysts of the steam-powered Industrial Revolution.

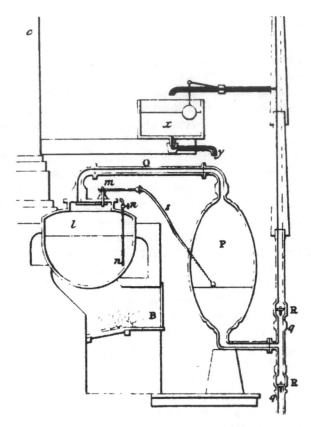

Fig. 3 A Savery steam engine, side view, 1702 (*Credit* Wikimedia, Thomas Savery)

Why Was the Steam Pump Invented?

The direct purpose of Savery's invention is clear. It was designed to solve the problem of pumping water out of mines. Its mode of operation—with no moving parts—means that we probably can't claim that he was trying to invent a steam engine, in the sense of Newcomen or Watt—i.e. an engine that produced rotary motion. However, Savery did show the potential of steam to produce useful work, and it did directly contribute to the later inventions of Newcomen and Watt. So there are two reasonable candidates for the problem. Either, *how to pump water*—which is the very specific problem he was addressing—or possibly the more general issue of *how to convert steam into useful work*? Either way, his invention was highly creative, as we will now see.

How Creative Was the Steam Pump?

Relevance and Effectiveness: Savery's steam pump scores 3.5 out of 4 in this category. It was technically appropriate and made correct use of the knowledge of heat, thermodynamics and mechanics available at the time. Indeed, on paper, or we might say *technically*, it's a good, clever solution to the problem. It fits within constraints ranging from cost—it was sufficiently affordable to be a commercially available device—through to the technology used. The one black mark in terms of relevance and effectiveness is its actual performance. In theory, it could solve the problem of pumping water out of mines, or of supplying water to towns, but in practice, it frequently encountered difficulties that could not always be overcome. On one occasion, for example, during a demonstration in London, the pump created steam at a pressure of 8–10 atmospheres but blew out the various joints on pipes and other parts of the pump. On another occasion, attempting to use the pump in a mine, the device blew itself apart and was abandoned. There were many instances where it was used successfully, but also many where it broke down, and we have to deduct some marks here because it cannot be regarded as fully, reliably effective.

Novelty: In many respects, the steam pump scores very well against the criterion of novelty. It is an excellent example of an invention that took known concepts—in this case, general principles of heat and steam, for example—and pushed these in a new, and very practical, direction. In this sense, it was both incrementally valuable but also pushed that incrementation towards wholly new pathways. In other words, if it is possible for something to be both incremental, and radical, the steam pump invented by Savery may be a good example. There are only two aspects of novelty that I feel were slightly less than perfect. Because of the difficulties in getting this device to function as desired, it becomes a little harder to see how it draws attention to shortcomings in other solutions. Think of the water-raising example that I discussed when we considered Al-Jazari's crankshaft and its applications. If we put the steam pump, and Al-Jazari's 'pump' side by side, we might have trouble showing that Savery's device was an improvement, given its unreliability. The second, and related, deficiency of the steam pump lies in how it helps the observer to see the benefits of changes—if it can't be made to work reliably, it becomes harder to experiment with it! Notwithstanding these relatively minor issues, it scores a strong 3.5 out of 4 for novelty.

Elegance: The minor problems that detract from the steam pump's effectiveness and novelty both may be traceable to a problem of elegance. We have already seen a connection between creativity characteristics like effectiveness and elegance, and I have explained that good solutions usually *look like* good solutions. The principal drag on Savery's device, which of course was rectified in later steam engines, was what we might describe as poor (relatively) execution. If the pump couldn't always withstand the pressures generated by the steam, for example, then there is some sort of mismatch between the idea and its execution, and one that leads to poor effectiveness. This is not to say that Savery or his technicians were careless, or deliberately set out to make an inelegant device, but the end result is the same. In fact, the deficiency in elegance was almost certainly a reflection of the device's high novelty—which is rather paradoxical! The more you push the boundaries of what is possible, or introduce radically new ideas, the more difficult it may be to execute such an idea elegantly, and therefore the weaker the effectiveness of that idea. Could it be that high effectiveness can't coexist with high novelty? In other words, is it impossible to achieve a perfect score on our scale of creativity? We'll try to explore this further, but for now, Savery's steam pump scores only 2.5 out of 4 on the criterion of elegance.

Genesis: Possibly reinforcing the point just made—can high effectiveness and high novelty coexist—I have scored Savery's steam pump at the maximum 4 out of 4 for genesis. The steam pump must score well across the board in this category. It clearly suggests a novel basis for further work, and indeed, later improvements and developments would bear this out. The steam pump also stimulates in the observer many ideas for solving unrelated problems. The possibilities opened up by the basic steam power concept demonstrated in this device would lead first to better, more reliable pumps, but also to locomotives, textile manufacture, electrical power generation, and more. The steam pump also changed the basic paradigm of water pumping from one using mechanical aids to lift the water, to one that used differences in pressure, and the atmosphere itself, to do the lifting. In all of this, the steam pump also set a fundamentally new benchmark for judging mechanical aids of many kinds. It broke existing paradigms, and injected a major disruptive element to the field of mechanics, and therefore scores very well for genesis.

Total: Thomas Savery's Steam Pump scores a strong, but not perfect, 13.5 out of 16 for creativity. If the technical means had existed at the time to execute it better, we would be looking at one of our strongest inventions. Nevertheless, it is clear that Savery's invention made an important contribution, and the true creativity is realised in the later improvements of Newcomen and Watt. The latter two inventors—especially James Watt—are better known than Savery, but we should not overlook his important contribution which heralded the coming Industrial Revolution.

The Age of Enlightenment (1685–1815): The *Business* of Invention

The Age of Enlightenment is perhaps the natural consequence of the preceding Age of Exploration. As our forebears began to exert some degree of control over the world around them, and as their understanding of that world developed and grew, it is no surprise that the kinds of problems that exercised our ancestors' minds moved from the more physiological and basic to the philosophical and advanced. Many of the upheavals of this era were associated with *reason*, and a general shift in interest away from systems premised on subservience and ignorance to systems premised on individual liberty and religious tolerance. A general attitude developed in this era that individuals should *dare to know*,[1] and the more our forebears dared to know, the more knowledge they gained, and the more problems they had to solve.

Our first innovation in this era is the Leyden jar. This early form of electrical battery gave scientists the first controllable source of electricity and was a vital necessary first step for the development of electricity as a viable source of power (that would ultimately replace steam). The second invention that we consider in this period is not a tangible entity, but a larger system—the Industrial Revolution. What this really means is the development of steam as a practical source of power. The critical innovation here was the ability to use steam power to produce rotational motion, and that had profound consequences. It meant not only that human activity, and human power, could be supplemented or replaced by a much greater form of power, but that the kind of work performed was changed. Our final innovation in the Age of Enlightenment is another medical invention—in this case, the invention of the smallpox vaccine. This is another good example of a shift in the kinds of problems that humankind could, and would, solve. Although it addresses a very basic, physiological need—preventing a deadly disease—the smallpox vaccine also shows how our ancestors could apply reason and logic, coupled with careful observation, to solve a problem that would also then address higher order needs. A society free

---

[1] First attributed to Roman poet Horace, in his *First Book of Letters* (20 BCE), the phrase was also used by German philosopher Immanuel Kant (1724–1804) in his essay *Answering the Question: What is Enlightenment?* (1784) and became a catchcry of the Enlightenment period.

© Springer Nature Singapore Pte Ltd. 2019
D. H. Cropley, *Homo Problematis Solvendis—Problem-solving Man*,
https://doi.org/10.1007/978-981-13-3101-5_8

from widespread, deadly illnesses can put itself into a position to consider higher needs of esteem and self-actualisation, in a way that a sick society cannot.

The Leyden Jar (1745)

Electricity is but yet a new agent for the arts and manufactures, and, doubtless, generations unborn will regard with interest this century, in which it has been first applied to the wants of mankind – Alfred Smee, English Polymath (1818–1877)

As the period known as the Enlightenment progressed, the knowledge that was discovered, and rediscovered, during the Renaissance began to bear fruit. This is the natural progression of the development of scientific knowledge. An initial period of observation and theory generation—often called *induction*—leads to a transitional phase in which we work out how to measure and understand the core properties of the things we observe, and the relationships between these properties. As this knowledge develops, we move into a *deductive* phase of theory testing—if I do A, then B will happen—that leads to the practical application of knowledge. The Leyden jar seems to be an example of how the transitional phase of knowledge development unfolds. Basic notions of electricity, developed through observation, then lead to, and require, mechanisms to delve more deeply, to understand what is going on, and how this mysterious phenomenon works. Without this, and the Leyden jar, the means to progress to the deductive, and applied, phase of knowledge could not proceed.

What Was Invented?

The Leyden Jar, named after the Dutch city of Leiden, which was home to one of the two scientists credited with its invention, is a simple form of electrical capacitor (or condenser), capable of storing an electrical charge. This breakthrough played a crucial role in the study of electricity creating, in effect, a controllable battery.

The device itself consists of a glass container, with internal and external metal coatings, and two terminals (or connectors). One of these terminals is in contact with the inner metal surface, projecting through the jar's stopper (so that it is insulated from the outer surface). The second terminal connects to the outer metal surface. In essence, therefore, the Leyden jar consists of two metal plates, separated by a thin, insulating layer of glass. Leyden jars were charged usually from a friction generator (for example, a rotating glass ball in contact with a woollen cloth, or the better known Van de Graaff Generator, that many readers will have experienced at school). Leyden jars are high voltage devices, capable of storing a charge in excess of 30,000 V (see Fig. 1 for an example of an array of Leyden jars creating a simple battery).

The Leyden jar appears to have been invented, almost simultaneously, by two men. One was the German priest and physicist Ewald Georg von Kleist, and the other was

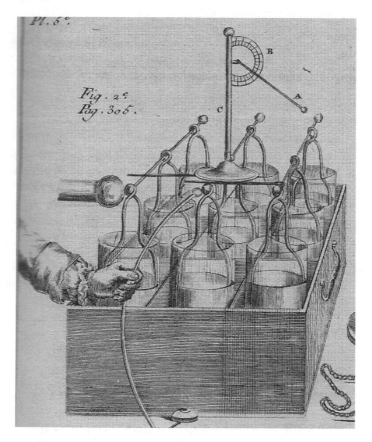

Fig. 1 An assembly of Leyden Jars to form an electric battery, 1773 (*Credit* Wikimedia, Benjamin Franklin)

the Dutch Professor Pieter van Musschenbroek (who worked at the University of Leiden at the time he developed his version of this device—hence the name). Both men were interested in the study of electrostatics (the study of stationary electrical charges or fields, in contrast to electrical currents). This had become popular in the second half of the seventeenth century after the invention of simple friction machines. Both were following up on the work of Georg Matthias Bose, who, by the age of 28, in 1738 had become the Professor of Natural Philosophy at the University of Wittenberg in Germany and was famous for his experiments in this field.

Regardless of who invented it, and there seems little reason not to credit both von Kleist and Musschenbroek, the Leyden jar was a watershed invention in the developments associated with the study of electricity. Once humans were able to store and use electrical charge at will, the door was open to its application in a range of different ways. In the next section, we will meet the First Industrial Revolution, which was the first great technological leap forward based on water, and especially steam power. It would take some time to develop further, but electricity would be the spark (pun intended) that ignited the Second Industrial Revolution beginning in around 1870, and the Leyden jar laid the foundation for this later advance.

Why Was the Leyden Jar Invented?

The Leyden jar is somewhat unusual, as far as inventions go because it did not address a problem that would have been considered very practical or important by most people. It was, nevertheless, vital to the development of later solutions that have had enormous impact. In order to create the electrical devices that are so widespread nowadays—lightbulbs, electrical motors, computers and so on—our ancestors first needed a means for isolating this elusive form of energy. Lightning strikes told our forebears that there was enormous power in electrical energy, and individuals occasionally experienced this in a small way through a static shock, but where did this come from, and how could it be controlled? This was the contribution of the Leyden jar—it solved the problem of *how to study electricity*.

How Creative Was the Leyden Jar?

Relevance and Effectiveness: The Leyden jar appears to be such an innocuous device—a glass jar with some metal-coated surfaces—and yet it was extremely effective in storing electrical charge. We are not talking about the kind of static electricity that gives you a small shock when you rub your feet on carpet (which might be around 5000 V). The Leyden jar, even in these early days, could store a charge estimated to be as much as 60,000 V, which was both sufficient to cause a much more painful shock, and also more useful to physicists. Incidentally, don't confuse these seemingly large numbers with the effect produced by current electricity (i.e. what you get from your wall socket). A 110 V supply can kill you because the current produced is sufficiently large, and it is the current that does the damage in an electric shock. The value of the Leyden jar was the *control* that it gave scientists. While this was a long way from powering electrical machines, it did provide the key resource needed to study electricity in a systematic and rigorous manner, and the Leyden jar, therefore, deserves a high score of 3.5 out of 4 for relevance and effectiveness.

Novelty: The Leyden jar scores well in this criterion. Although this innovation could be said to originate with a rather serendipitous discovery, both von Kleist and Musschenbroek were attempting to replicate, and improve on, Bose's earlier work with water, alcohol and electricity. Their work highlighted both the shortcomings of the previous work and clarified how and why changes might improve on what was known. Once some of the basic electrical properties of the materials had been worked out, basic criteria of novelty were able to be fulfilled—for example, the Leyden jar showed how the basic device could be extended into a larger array of jars. Also, with the basic device perfected, improvements were rapid, and over the next couple of years, many refinements were made to the core principle of the electrical battery. For these reasons, I give the Leyden jar a score of 3.5 out of 4 for novelty.

Elegance: Although rather pleasing in appearance, and carefully constructed, the Leyden jar had some weaknesses on this criterion. In addition to the fact that its working principles would not have been apparent to most observers (how does this empty jar hold this invisible electricity?), many of the early researchers inadvertently shocked themselves when handling these devices. Even the early electrical researchers were not entirely sure how or why the devices did what they did. A more elegant device might have anticipated these aspects of the invention and incorporated some safety features that prevented accidental shocks. We can again link elegance to effectiveness—the Leyden jar was both easy to charge with electricity, and easy to discharge. So much so that it often shocked the user! I give it 3 out of 4 for elegance.

Genesis: The Leyden jar established a paradigm, rather than breaking one. Prior to its invention, even the Ancient Greeks knew that rubbing a piece of amber would cause it to attract small particles. The problem was capturing this charge in a controllable manner. With the invention of the Leyden jar, scientists had the basis for an entirely new discipline. New problems could be explored and new norms were established. Perhaps the only weakness in this criterion is the fact that the principles behind this device were so new that some scientists at the time may not have been sure how to make use of it! I have given it a score of 3.5 out of 4 for genesis.

Total: The Leyden jar's overall score for creativity is 13.5 out of 16. This places it into the *very high* range. There are only minor weaknesses in each category, perhaps not surprising for such a groundbreaking device. Given that fact, the main impediment to a higher score is execution. Not only was the Leyden jar surprising, but it was still a little mysterious, even for the experts. If the underpinning concepts had been better understood, we can imagine a higher score, close to the top of our list of inventions.

The Industrial Revolution (c1760–1830)

…a revolution which at the same time changed the whole of civil society – Friedrich Engels, German Philosopher (1820–1895)

The Industrial Revolution is different from most of our inventions because it is not a single, identifiable or tangible artefact. Instead, it represents a more diffuse idea that is probably better described as a process or even a system. That raises a few issues for our analysis that are worth discussing. The first is whether or not an *intangible* idea, for example, can be considered, and indeed assessed, as creative. While it is certainly easier to rate and discuss the creativity of a tangible artefact—a Spinning Wheel, a crane and a pendulum clock—there is nothing that stops us from applying the same criteria to other, often abstract, things. In fact, as I hope I have made clear in earlier discussions, what we are exploring and assessing in this book are not, in essence, tangible artefacts, but solutions to problems! A solution to a problem can also be a process—a way of doing something—such as Henry Ford's introduction of the production line for manufacturing automobiles. You can't hold the process in your hand, or touch it, but it exists nevertheless, and it is a solution to the problem of how to mass produce artefacts. Equally, a process, as a solution to a problem, can be relevant and effective, novel, and elegant, and it can possess the quality genesis, just the same as any tangible artefact.

Similarly, a solution to a problem can also be a system. A good definition of a system is that it is a combination of interacting elements forming a complex, unitary whole. We tend to see examples of systems more in recent times. To some extent, at least in technological terms, complex systems are a modern phenomenon resulting from the fact that it is much more feasible nowadays to create complex combinations of, for example, hardware, software and people, in ways that weren't possible in bygone eras. A good example of a complex system is a modern hospital. These now represent incredibly complex combinations of people (doctors, nurses, administrators and even patients), hardware (the buildings, heating, lighting, computers, medical instruments and equipment) and software (in the information systems, the communication systems and health records). In addition, these elements are all strongly interconnected, which is part of what makes these systems so complex. Like processes, systems can also be judged for their creativity, even though they also can't be held in your hand, and even though parts, at least, are intangible.

Finally, a solution to a problem can also take the form of a service. What I mean here is an organised system of labour and material aids used to satisfy defined needs. So examples of services are things like a bank account, a retirement plan, babysitting or even pizza delivery. Typically, these are also intangible solutions to problems. You can't physically hold your bank account in your hand, but nobody would deny that it exists. Like our other alternatives to tangible artefacts, services can also be assessed for their creativity. How well do they solve a defined problem, and how novel, elegant and so forth are they?

Back to the Industrial Revolution—I decided to include it in this catalogue of innovations for two reasons. First, it takes us into some slightly different territories, showing the point I have just been discussing. It is not a tangible product but is a more diffuse, intangible process or system that serves as a solution to a problem. Second, because it is so pivotal to the history of technology, and is itself a technological solution, it seemed like an intangible solution that we could not ignore. I'm now declaring that I will treat it as a *system*—a combination of interacting elements forming a complex, unitary whole—and an energy-handling system at that, and we will assess its creativity against that basic definition. Before we do that, however, we need a quick description of this system of interest.

What Was Invented?

Some experts define the Industrial Revolution in terms of political, philosophical and even social changes. However, in this book, I will define it in *technological*

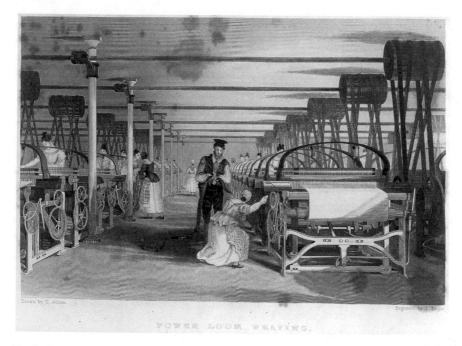

Fig. 2 Power loom weaving, early 1800s (*Credit* Wellcome Collection)

terms. In particular, I argue that the Industrial Revolution was a function of the development of steam power. Not the relatively crude attempts of Thomas Savery that we discussed in the previous chapter, and not even the subsequent refinements of Thomas Newcomen—whose engine produced reciprocating motion in the first decades of the 1700s—but in terms of the further innovations of James Watt in the second half of the century.

The key difference and development between these three applications of steam power lies in the kind of work that these different engines were capable of producing. Savery's steam pump, with no moving parts, lifted water and had application in keeping mines free of flooding. Newcomen vastly improved the efficiency of this process and produced a more useful, but still limited, reciprocating (i.e. up and down, piston-like motion). This was also primarily used to pump water in mines. The real advance that would drive the technological changes of the Industrial Revolution—and James Watt's major contribution—was the ability to use steam power to produce *rotational* motion. The reason this advance was so important is that it allowed the application of steam power to mills (see, for example, Fig. 2), grinding, turning lathes and other functions such as raising coal and iron ore. The transition from Savery and Newcomen to Watt was a change from steam power assisting in the *primary production* of raw materials, to steam power assisting in the production of *value-added* goods. What the Industrial Revolution represents is the massive impact of steam-powered *mechanisation* on manufacturing.

Why Was the Industrial Revolution Invented?

The term *invented* doesn't quite fit our context here. We have used the term to mean something that was deliberately designed, by humans, to solve a defined problem. On the other hand, discovered—implying that we merely gain knowledge of something, often unintentionally—isn't right either. Nobody discovered the Industrial Revolution. I'll stick to invented, but we have to see that in more of a systems sense, in the same way, that the Industrial Revolution itself is a system of interacting elements (people, machines, steam power and so on). One other useful term here is the concept of emergent properties. These are characteristics of complex systems that are apparent only when all of the pieces of the system are working together. Thus, the ability of a bicycle to act as a form of transport is an emergent property of the wheels, frame, pedals, chain and rider all joined together. The reason that this concept helps us to understand the Industrial Revolution is that, like a bicycle, we can consider the Industrial Revolution as the deliberate result of the interaction of steam engines, people, cotton mills, factories and so forth. Most importantly, it's clear that none of the things came about by accident. Each was designed, and therefore it's reasonable to say that the system—the Industrial Revolution—was also a deliberately created outcome.

Given that viewpoint, what problem or need did it address? A reasonable problem statement might be—*how to mechanise society*. In more everyday terms, it's not hard to imagine that the architects of the Industrial Revolution were motivated by the question of how they could apply the power of machines to relieve humankind of the drudgery of many of our daily activities and satisfy a growing demand for manufactured goods. Although there is no doubt that they succeeded, they failed to anticipate one other property of complex systems—the problem of undesired emergent properties. We often refer to these as *unintended consequences*.

How Creative Was the Industrial Revolution?

Relevance and Effectiveness: Considering the Industrial Revolution more specifically as a system for the mechanisation of manufacturing, I find it hard to go past a maximum score for this criterion. A key characteristic of this innovation is the way that it represents a set of changes all based on the collective advancement of scientific knowledge and its practical application. The solution system in this case—namely, the steam-powered mechanisation of manufacturing—did what it was supposed to do in a variety of different ways, and did so within the existing constraints, or found a way around those constraints. Unlike our other innovations, we allow ourselves here to consider the solution not only as a system but also as one that evolved during the time period. I, therefore, give it 4 out of 4 for relevance and effectiveness.

Novelty: Against all the specific indicators of novelty, both in the way that the Industrial Revolution helps to define and illustrate the problem at hand and in the way that it sheds new light on that problem (the problem being the mechanisation of manufacturing), this system innovation scores very highly. The application of steam power to the generation of rotational motion encapsulates the fundamental problem that was waiting to be solved. The problem was not simply to find ways to use machines to make tasks easier for humans. Levers, for example, did this but did not give rise to the massive technological, social and political changes of this era. The problem was to find ways to transform (and not merely augment) human activity, especially more complex human activities like the manufacture of sophisticated tools, instruments, goods and services. The Industrial Revolution fundamentally changes humankind's understanding of its relationship to technology and, in this way, scores 4 out of 4 for novelty.

Elegance: The only blemish, from a creativity point of view, for the Industrial Revolution lies in the criterion of elegance. Recall that elegance encapsulates the quality of execution of the innovation. This is both in a more literal, technical sense—i.e. how well made is the innovation—and also in a more aesthetic sense. In other words, is the solution pleasing and neat to the observer? This brings into consideration factors like the impact of the innovation on its surrounds and allows us to consider how it interacts with its environment. One of the characteristics of systems is the fact that no system is ever really closed in the sense that it is wholly self-contained. No matter where we draw the boundary of the system that is the Industrial Revolution, we find interactions with whatever lies outside of that boundary. Unfortunately, also, systems give rise not only to the desired properties—those that we build into the system to solve the problem at hand—but also undesired properties, not least factors like pollution. This is the blemish on the Industrial Revolution. We cannot deny that it created wider and undesirable effects, both environmental and social. It gave rise to changes to the landscape, pollution and waste, and may even be the catalyst of our modern thirst for energy, without (at least until recently) much concern about the impact on the environment. The social impact of the Industrial Revolution was also widespread, and frequently negative. People left traditional jobs on the land and were often employed in horrific conditions, with severe impacts on health and social relationships. Children were exploited in factories, and cities became crowded, unhygienic and polluted. If you wonder why I am only deducting half a point for this inelegance, it is because our primary focus has to remain inside the system boundary—the solution itself. However, the score of 3.5 for elegance cautions us to remember that many innovations, especially more complex system innovations, do typically have wider external impacts that we should always keep an eye on.

Genesis: The Industrial Revolution was a game changer. Whereas novelty is mainly concerned with the nature of the solution itself and how it defines and illustrates the problem, genesis is about how the solution alters our understanding of the problem. Here, our chief interest is the fact that this innovation completely replaced the previous paradigm of human labour, mechanical and other aids. It was not simply a slight improvement—a longer lever or an extra pulley—but was a completely new concept, both in terms of how assistance was supplied (steam power) and where that assistance was applied (to manufacturing, rather than primary production). There would be further game changers in the future, and we are now living in the era of the Fourth Industrial Revolution. However, for at least 100 years, this First Industrial Revolution made its impact felt across the world. I score it 4 out of 4 for genesis.

Total: The Industrial Revolution scores a near-perfect 15.5 out of 16 on our scale of creativity. As we have seen, the only slight negative was the inelegance of its wider impact. While that is not trivial, I explained that our focus in these analyses has always been the innovation itself. Even though this innovation scores very highly, we also see that this did not mean that it represented an end point in a process of innovation. Some one hundred years later, the world would experience a Second Industrial Revolution, with the advent of electricity, mass production and the use of assembly lines. Another one hundred or so years after that, a Third Industrial Revolution began, thanks to the introduction of computers and widespread automation. Now, in the early twenty-first century, we are at the start of what is referred to as the Fourth Industrial Revolution, or Industry 4.0, characterised by the introduction, in manufacturing, of big data, artificial intelligence, cloud computing and similar technologies to create what are often called cyber-physical systems.

The Smallpox Vaccine (1796)

...when that day comes, there will be no more smallpox – Edward Jenner, English Scientist (1749–1823)

The invention of the smallpox vaccine not only saved millions of lives, and improved the quality of millions more, but helped to drive a transition in medicine that would eventually save millions more. Although still mired in an era that did not fully understand infection, germs and other related concepts, Jenner's work demonstrated that disease was not beyond the control of humankind to treat. It helped to cement a growing confidence in our ancestors that the world was not a mysterious and predetermined mechanism, unfolding according to some incomprehensible plan. In the same way, that energy could be controlled and manipulated—as evidenced by the advances in steam power—health and wellness could also be subjected to scientific investigation and manipulation. The smallpox vaccine gives us an example of a material handling system, in this case, handling substances, and creating effects, on a microscopic scale.

What Was Invented?

Edward Jenner was an English physician and scientist of the Enlightenment era. As the creator of the world's first vaccine—a term originally referring only to smallpox, but now used to refer to any similar protective process—he is now often credited with saving more lives than any other person in history.

Jenner developed the smallpox vaccine based on his observation that people who had previously been infected with cowpox—a similar, but much milder disease—did not contract smallpox. This fact was a common belief amongst farmers and other who worked with cows—for example, milkmaids—but at the time had no formal basis in science.

Jenner was not the first to invent the broader concept of deliberately protecting individuals from smallpox by some artificial means. A less sophisticated method——inoculation or variolation—had been practiced in a number of cultures from as early as the tenth century (for example, in China). This process might involve powdering the smallpox scabs from someone infected with the disease and blowing this powder up the nose of the patient! Typically, this would induce a mild form of smallpox, after which the patient was then immune to further infection.

Jenner's breakthrough was to establish a formal link between cowpox and smallpox and to develop a process for inoculating patients against cowpox, which then created an immunity to smallpox. In other words, he removed the need for a patient to be infected with *any* form of smallpox (mild or otherwise), in order to be rendered immune from that disease. In doing so, he systematised the existing heuristic knowledge linking cowpox immunity to smallpox immunity and rendered the process of vaccination safe and simple.

Jenner's smallpox vaccination process involved transferring pus from cowpox sores on one patient to a small wound on another patient (see Fig. 3). The method, with modern refinements in the production, storage and application of the treatment, proved so effective that a worldwide campaign, between 1967 and 1979, was able to officially declare smallpox eradicated by 1980. Readers born from 1980 onwards therefore do not possess the characteristic scar on their upper arm—a result of the smallpox vaccination—that older readers will be familiar with.

One interesting fact to keep in mind is that, although declared eradicated in 1980, two countries (the US and Russia) do retain stocks of the smallpox vaccine. It is unclear if the vaccine confers permanent immunity, and there always remains the risk of a future outbreak, either naturally or possibly as an act of terrorism.

Why Was the Smallpox Vaccine Invented?

The driving problem behind the smallpox vaccine is not hard to guess. Smallpox was a highly infectious, disfiguring and often deadly disease. Jenner's research and his vaccine were not happy accidents, and the smallpox vaccine represents a very deliberate attack on a very specific problem. The underlying drivers may be linked to basic, physiological needs, as well as higher psychological needs, and may even indirectly relate to the fulfilment of the highest needs in Maslow's hierarchy. However, the direct problem that Jenner solved was *how to prevent smallpox infection.*

HILLEMACHER (E. E.). H. C. *Edward Jenner faisant ses premières expériences de vaccine.*
Edward Jenner's first experiences with vaccine lymph.

Fig. 3 A young Edward Jenner vaccinates an anxious-looking child (*Credit* Wellcome Collection, E. E. Hillemacher)

How Creative Was the Smallpox Vaccine?

Relevance and Effectiveness: While Jenner's vaccine certainly worked, it is noteworthy that nobody at the time really understood *why* it worked. Pasteur's now universally accepted *Germ Theory*, for example, was still decades in the future. Jenner's invention, which was certainly deliberate and systematic, nevertheless involved a slice of luck that just knocks a little gloss off relevance and effectiveness. If we had been present at the time, and facing a severe, certainly disfiguring, and possibly fatal smallpox infection, I'm sure we would have been delighted to be treated by Jenner. However, he would not have been able to guarantee that his vaccine would work, or that it would produce no undesired side effects. Jenner also believed that it would confer lifelong protection, which was not the case. For these reasons, I give it 3.5 out of 4.

Novelty: The smallpox vaccine was a significant step forward in showing how existing methods could be improved. Removing the risky step of infecting patients with smallpox, and relying only on cowpox to confer immunity, not only made the process safer but also paved the way for a deeper understanding of diseases and disease prevention. It began a process of moving from intuition to evidence-based disease prevention. At the same time, it was *not* a radically new approach. Like many of our cases, it was an incremental step forward, and one that built on a foundation of previous, informal, knowledge. For these reasons, it can be said to have a degree of novelty but remains on the same curve (think *diminishing returns*), rather than stepping onto a new curve. I give the smallpox vaccination 3 out of 4 for novelty.

Elegance: It is difficult to describe Jenner's smallpox vaccine as highly elegant in the sense of being a complete, well-rounded and fully formed solution. In many ways, it was a rather crude, almost *backyard* solution. Jenner's initial test involved the rather unsavoury process of collecting pus from sores on a milkmaid—Sarah Nelmes—who had been infected with cowpox. This was transferred to an 8 year-old boy—James Phipps—by scratching his arm and applying the pus to this small wound. No sterile needles, infection control or laboratory procedures in Jenner's day! As Jenner expected, the boy fell ill with cowpox, experiencing only a mild fever and producing a cowpox sore at the site of the inoculation. Phipps recovered after a few days, and Jenner then waited 2 months before attempting the final and equally unsavoury step. He collected scabs from smallpox sores on another patient and used these to inoculate Phipps. As Jenner had hypothesised, the cowpox inoculation had successfully conferred smallpox immunity on Phipps, and the boy showed no signs of infection from the smallpox that had been applied to scratches on both of his arms. Notwithstanding Jenner's success, the fact remains that there were many unknowns in this process, and Jenner himself experienced some difficulties, for example, in distinguishing between *true* cowpox and a so-called *spurious* form, that either failed to produce smallpox immunity or caused an adverse reaction. In simple terms, Jenner's vaccine was highly experimental—incomplete and crudely executed—and it is difficult to score higher than 2.5 out of 4, despite its important status in disease prevention.

Genesis: Perhaps *because of* the smallpox vaccine's relative inelegance—its messy, experimental nature—it scores quite well for genesis. The vaccine demonstrated a novel basis for further work. Not only could this concept be refined for smallpox, but it might open pathways for protecting against other diseases. It suggested a new way of looking at diseases, and also highlighted weaknesses in existing processes, such as direct inoculation with smallpox. Jenner's concept immediately rendered previous approaches to the protection of individuals against smallpox obsolete—both the informal process whereby milkmaids were thought to derive immunity through their contact with cowpox and the more deliberate methods that relied on infected material is taken from smallpox patients. I give Jenner's invention a score of 3 out of 4 for genesis.

Total: The smallpox vaccine developed by Edward Jenner scores a total of 12 out of 16 on our scale. This places it at the upper end of the *high* range for creativity. Two weaknesses stand out that hold it back from a higher score. First, it was not only an incremental solution—there had been some crude understanding of the mechanism for a considerable time—but Jenner was not even the first to try and use cowpox as a means of protecting individuals against smallpox. Thus, the innovation loses some points for novelty. In addition, as I have already indicated, the elegance of the solution also leaves quite a lot to be desired. Nevertheless, it is perhaps an indication of the different ways that we assess risk that, in relation to the potential reward, we might be unwilling to set foot on a bridge that scored only 2.5 for elegance, but would willingly accept vaccination against a deadly disease with the same relative lack of completeness and good execution!

The Romantic Period (1800–1900): Accelerating Change

The Romantic Period—running roughly from the late 1700s to the late 1800s (or approximately, the nineteenth century in its entirety)—is probably characterised best by the consequences of industrialisation. This era has also been referred to as *The Great Divergence*[1] and while there is a debate about exactly when it began, there is little disagreement on the consequences. This was a time of rapid and significant change, facilitated by ever-increasing technological prowess, and is therefore most evident in the so-called Western World (i.e. Western Europe and North America). Electricity, as a usable utility, emerged in this period, along with many of the resulting innovations that are fundamental to modern life—refrigeration, railways, telegraphic communications, telephones, automobiles and so on. In many respects, the Romantic Period saw the fruits of the Industrial Revolution emerge, though not without many social, economic and environmental costs.

The first invention that we consider in this era is the humble *velocipede*. In essence, the first bicycle, this device was more than just a plaything. It arose from a very practical problem—lack of feed for horses—and demonstrated humankind's increasing ability to manufacture almost any kind of solution to any problem. What I mean here is that in the previous eras, our ancestors were limited by what they could produce. They might imagine a great solution to a problem, but not be able to make it a reality. The velocipede, however, represents the beginnings of an era in which our capacity to *realise* our ideas was matching our capacity to think them up. The second invention in this period is another system. In this case, modern *sewerage systems* used to treat the waste products that followed closely behind industrialisation, and the rise of densely populated cities. This is a material-handling system, in a very concrete sense, and again shows a change in how our ancestors responded to problems of any level. We have an unanticipated problem of human waste and pollution in cities. We have the means to devise and implement a successful solution. Not only that, but our ancestors at this time also began to cast their minds forward, realising that the solution is so complex that it cannot be replaced easily. Therefore, it would be wise

---

[1]See the book *The Great Divergence* by Kenneth Pomeranz (2000).

© Springer Nature Singapore Pte Ltd. 2019
D. H. Cropley, *Homo Problematis Solvendis—Problem-solving Man*,
https://doi.org/10.1007/978-981-13-3101-5_9

to devise a solution not only for the problem as it is now, but the problem as it might be in another 100 years. Our final innovation in this era is one that has become synonymous with the concept of *ideas*—the *lightbulb*. Edison's invention is fascinating not only as a solution to the problem of providing electrical lighting to homes and businesses, with all the benefits that this entails, but also an exemplar of the *business* of invention. This period was also one in which people began to realise that there was money to be made in finding solutions to humankind's needs.

The Velocipede (1817)

> Every time I see an adult on a bicycle, I no longer despair for the future of the human race
> – H. G. Wells, English Author (1866–1946)

The first invention that we will consider in the Romantic era is a seemingly very simple example of an energy-handing system. The Velocipede, or Draisine, is a very close relative to our familiar and modern bicycle, minus the pedals. As simple as it looks, this invention had a far-reaching impact on a problem theme that has occupied humankind for much of our existence. In terms of Maslow's hierarchy of needs, we have previously discussed the place of devices such as oars. We noted that they did not necessarily directly satisfy a core need, but that they supplemented the solution of other basic needs, and possibly also contributed to the solution of psychological needs. The same appears to be the case with this forerunner of the modern bicycle. The fact is nobody needed this to satisfy their need for food, water, warmth or rest and probably not for security or safety. And yet, it may have made the satisfaction of some of those needs a little easier. It may also have contributed to the satisfaction of needs such as the psychological need for belongingness and love. It certainly may have satisfied needs associated with feelings of prestige, and we know that one of its nicknames was the *dandy horse*, suggesting that there was some sense that it was considered rather fancy in its day. Whatever place it occupies as a solution in a hierarchy of needs, it has been extremely influential as a stimulus for the development of other augmented forms of transport. It's even curious that the modern impetus for motor vehicles and other forms of transport may be simply the need to show off to our fellow humans!

What Was Invented?

The Draisine, or Draisienne, was the invention of German Karl Freiherr von Drais. Although born a Baron (German: Freiherr), he dropped his title and the "von" from his name in 1849, so we'll refer to him as Karl Drais. He invented the first bicycle—also frequently referred to as the Velocipede—and in doing so started a long progression of mechanised forms of personal transport. Not only that, but it also was the beginning

Fig. 1 One of the first Draisines built by Karl von Drais (*Credit* Wikimedia)

of the end for horses as a means of land transport, or as the source of power for forms of land transport.

The Velocipede consisted of a light wooden frame supporting two wooden wheels. The rider sat astride this frame on a saddle (… it was a bicycle, and I probably don't need to explain it) and rather than pedalling, as we do on its modern equivalent, the rider pushed along the ground with his or her feet. Although it lacks the modern refinements of a more comfortable saddle, pneumatic tires, pedals and gears, we can see (Fig. 1) that it is instantly recognisable, and that any modern person would immediately understand how to use this device. The Velocipede also included a friction brake on the rear wheel, and a rather elaborate steering mechanism, and the Velocipede was in every sense of the modern word, a bicycle.

Although not directly a quality or property of the device itself, there were some interesting knock-on effects of the Velocipede. As readers can imagine, in the early 1800s, there was not the system of good quality, paved roads that we are familiar with today. Velocipede riders therefore first took to the highly uneven, muddy and rutted roads of the time. Not only did these make using the Velocipede rather difficult, but it was a dirty and uncomfortable ride. Riders therefore soon began riding on the better quality footpaths of cities in Europe. Unfortunately, this quickly led to problems because the Velocipede was seen as a hazard to pedestrians, capable of racing along at speeds of as much as three times a fast walk. It was soon banned in countries such as Germany, Great Britain and the United States, and was soon out of fashion. Although this might have been the end of the story, like many inventions, once introduced, it was hard to uninvent, and incremental improvements to the device, as well as improvements to roads and other infrastructure, ensured that bicycles and motorised forms of horseless transport would eventually take over.

Karl Drais, unfortunately, never profited from his invention. To begin with, he was unable to benefit due to his status as a civil servant in the German state of Baden. Later, Drais became embroiled in revolutionary politics (the reason he renounced his title) and ultimately had his pension confiscated by the Prussian authorities, and he died penniless in 1851.

Why Was the Velocipede Invented?

Some say that the stimulus for Drais's invention was the very specific need to find a form of transport that did not rely on horses. Not for reasons of animal welfare, or efficiency, but for the very specific reason that poor weather, and a volcanic eruption in Indonesia in 1815, had led to a series of poor harvests in Europe and a shortage of oats. This was putting pressure on the cost of horse-based transport, and Drais is said to have been seeking a solution to the problem of horseless (and therefore oat-free) transport. This example is also a perfect illustration of change as a driver of innovation. Climate change caused poor harvests. Poor harvests drove up the cost of feeding horses. Horses were widely used for transport. It did not take long for the *market pull* to define a new problem—how to free transport from its dependency on hungry horses. Drais responded directly to this new need with his innovative form of horseless, personal transportation.

How Creative Was the Velocipede?

Relevance and Effectiveness: This first example of a practical, steerable, human-powered, and widely used bicycle demonstrated a high level of relevance and effectiveness. The goal of Karl Drais, the inventor of this device, was to find a practical replacement for the horse as a means of transport. While our modern experience shows us that the bicycle would go through many further developments over the following 200 years, Drais succeeded in producing a design that was immediately practical. It is reported that it could be propelled along by its rider at speeds of up to 19 km/h and included many features that are familiar in today's designs. I give the Velocipede a score of 4 out of 4 for this criterion, reflecting its immediate impact and practicality.

Novelty: The simplest way to think of the criterion novelty is to ask how surprising the invention would have been to people seeing it for the first time. It is not hard to imagine that individuals in 1817 must have been amazed to see Drais's device! This is one of the few examples in our collection of inventions that must have emerged apparently out of the blue. There was no comparable invention that this replaced! The questions we must ask of the Velocipede include the extent to which it demonstrates shortcomings in other solutions. In this case, our comparison is to another popular transportation device of the day—the horse. In fact, we know that Drais was looking for an alternative to the horse, precisely because of one key weakness of horses: they need to be fed! The bicycle immediately reinforces this weakness, and shows how a device that does not have this weakness offers substantial advantages. Similarly in this category, we must ask to what extent an invention extends the known in a new direction. The bicycle did not use completely novel technology, but instead, drew in the existing knowledge—of wheels, bearings and techniques of metal- and woodworking—to create a new system with a new emergent property. In doing so, it offered a radically new approach to transport, and must be given a score of 4 out of 4 in this category.

Elegance: The thing that is possibly most striking about the Drais Velocipede is how well made it is. It reflects a high degree of craftsmanship and thoughtfulness in the design, with genuine attempts to offer some degree of comfort and operability to the rider. For example, it has a padded saddle, a rear brake and a steering mechanism that is designed to have a degree of self-centring in order to improve the usability of the device. It is also a device the purpose and method of operation of which is readily obvious. It would not have taken anyone familiar with riding a horse too long to understand where to sit, and how to propel the Velocipede. All of these factors, together with its quality of construction, and the clarity and simplicity of the design suggest a solution that is highly elegant, and that therefore also benefits from an improved effectiveness as a result. Even today, it looks neat, well done, and can be seen as a well-executed solution to the human-powered transport problem. I give it 4 out of 4 for this criterion.

Genesis: The most enduring thing about the Velocipede is how it broke an ancient transport paradigm. In doing so, it also succeeds admirably in fulfilling the qualities of genesis. The Velocipede quickly opened up new ideas for further work on alternatives to horses, and human-powered transport. It did not take long for improvements to the Velocipede to emerge, including, for example, pedals and a chain for improved propulsion. The device also set the scene for solving other, unrelated problems. For example, the problem of sharing footpaths across a variety of different modes of transport, and even the problem of wear and tear on boots (it was noted, at the time, that riders' boots wore out much more quickly than normal when using the Velocipede). Through these, and similar qualities, and ultimately by opening up an entirely new conceptualisation of transport, the Velocipede scores 4 out of 4 for genesis.

Total: It has probably not escaped the notice of readers that, up to this point in our catalogue of inventions, the highest score we have seen is a 15.5/16 for the Industrial Revolution, first invented in 1760. Here, however, our first maximum score was 16 out of 16! It may seem rather surprising that something as innocuous as a bicycle achieves this distinction, but careful reflection will, I hope, show why this is justified. To understand the creativity of anything, we ask four simple questions. First, does it work? In other words, does it solve the problem it was intended to solve? In the case of this first bicycle, we have a clear "yes". Second, is it new, original or surprising? Again, and placing ourselves in the shoes of someone in the early 1800s, the answer must be a resounding "yes"! Third, is the solution well executed? Again, the Velocipede leaves no doubt that this was the case, and even to a modern eye, looks like a well-constructed solution. Finally, does it break the existing paradigm? Again, this first bicycle gets a clear "yes" for this question. With the benefit of hindsight, we can see this very clearly in the subsequent development of the solution to what it is today. The bicycle jumped off an existing curve, at a point somewhere at or beyond the point of diminishing returns, and started a new curve, with plenty of room left for improvement. So far, in our exploration of the history of *Homo problematis solvendis*, we seem to have found the prime example of humankind's ability to solve practical problems.

Sewerage Systems (c1850)

[The principle, in building a sewer system, was]…of diverting the cause of the mischief to a locality where it can do no mischief – Joseph Bazalgette, English Civil Engineer (1819–1891)

What we often refer to as *sanitation*—systems and devices that handle human waste—in fact, existed long before the Romantic Period. These *material-handling systems*—in this case, I think that needs no further explanation—perform a vital, but largely unseen function, without which our modern cities would be impossible to live in.

We have excellent archaeological evidence of systems used in ancient Rome. If you have been to archaeological sites such as Pompei, near the modern city of Naples, you will have seen examples of quite sophisticated toilets that were typical of those developed and used in ancient Rome (Pompei was destroyed by the eruption of Vesuvius in 79 CE, so these systems were already fairly sophisticated some 2000 years ago). Indeed, these systems were quite modern, and used running water to carry away human waste. However, sanitation systems took a great leap forward in the middle of the nineteenth century, prompted by changes in society that had not previously been experienced. It was the growing populations of great cities such as London, driven by the changes stemming from the Industrial Revolution, which tipped the sanitation balance, and created an overwhelming and urgent need for a solution to the problem of processing our poo.

What Was Invented?

Even today, the Roman influence on public sanitation remains evident. In Paris, for example the famous *pissoirs*—public urinals—were known more correctly as *vespasiennes*, after the first-century Roman emperor Vespasian. These public conveniences reached their peak in the 1930s when there were more than 1200 in that city.

However, our focus here is the systems developed first in European cities in the mid-nineteenth century, in response to the health problems associated with urbanisation, growing populations and the contamination of fresh water supplies. The most famous of these may be the sewerage system developed in London, as a direct response to the crisis that first arose in the early 1830s.

Cholera was the catalyst for this invention. In 1831, Britain suffered an epidemic that killed some 6000 people. Cholera, as we now know, is an infection of the small intestine by a bacterium (*Vibrio cholerae*) that causes diarrhoea, vomiting and therefore dehydration and electrolyte imbalance. The symptoms may begin as soon as 2 h after ingesting the bacterium, usually through water that has been contaminated with human faeces containing the offending *Vibrio cholerae*. Obviously, untreated, it can lead to death. Unfortunately, the prevailing medical knowledge of the early 1830s had not yet embraced the so-called *germ theory of disease*, which holds that diseases are caused by microorganisms invading and multiplying in the human host. This fundamental lack of understanding of the cause meant that no measures were taken to address the underlying contamination of drinking water, and little in the way of effective treatment was given to people suffering from the infection. The most commonly held, and incorrect, theory of the cause of Cholera, and other infections, was the *miasma* (Greek: pollution) theory. This ascribed the cause to *bad airs* (hence the Latin term, *malaria*), or *night airs*, that were thought to emanate from rotting organic matter, so that preventative efforts usually focused on getting rid of the unpleasant odours that were thought to be the cause. Sulphur, for example was often burned in houses because this was thought to drive out the miasma.

The Cholera epidemics in London (a second severe outbreak occurred in 1848 and another in 1854) did, however, have an effect beyond the disease itself. John Snow, a physician in London at the time, made the connection between Cholera and the water supply, noting higher rates of the disease surrounding certain public water pumps in the city. He identified one, in particular, the Broad Street Pump, which he was able to link to a cluster of infections. It was through his pioneering work that the connection between Cholera and contaminated drinking water was established. This did not immediately lead to the adoption of the germ theory of diseases, but it did lead to a dramatic change in how human waste was treated in the growing, and increasingly crowded, City of London.

The innovation that we are interested in, therefore, was the sewerage system that was built with the express purpose of solving the problem of an accumulation of human waste that was filling the city, contaminating the river Thames, and therefore affecting the city's supply of clean drinking water. The new system was built under the supervision of Joseph Bazalgette, Chief Engineer of the London Metropolitan Board of Works, and was, in every sense of the word, a *system* solution. The fact is, there were sewers already in London at the time, but they had been designed to clear stromwater, and not to tackle high volumes of human waste. The system that Bazalgette created consisted of a network of 132 km of underground brick-built (see Fig. 2) main sewers, linked to more than 1800 km of street sewers that collected the raw sewage that would otherwise flow into the river. The system also incorporated pumping stations, and ultimately removed all the offending waste from

Fig. 2 Photograph of sewer tunnels at Wick Lane, East London, 1859 (*Credit* Wikimedia)

London, expelling it into the river far downstream (it's not clear what effect this had on people living in these areas!). Bazalgette also designed in a wide margin into the system, calculating generous allowances for the production of waste, and then doubling these, such that the system is still able to cope with the volumes of waste in the modern city. Ultimately, of course, it solved the problem of Cholera (and other waterborne diseases such as Typhoid) for the citizens of London, dramatically cutting these infections simply by removing (or perhaps we should simply say *displacing*) the source of infection.

Why Were Sewerage Systems Invented?

The obvious problem that led directly to the development of London's sewerage system was the question of how to dispose of human waste. However, a strong case can also be made that the more immediate problem was how to prevent Cholera, or even, how to avoid contaminating drinking water. These are all closely related, and to a large extent, solving one would solve the others.

How Creative Were Sewerage Systems?

Relevance and Effectiveness: As seems to be the case with most of our inventions, and probably not surprisingly, sewerage systems of the kind widely implemented in Europe and North America in the nineteenth-century score well for *fitness for purpose*. If they didn't, we wouldn't be considering them, and they probably would have been replaced with something that *did* work, and solved the problem(s) in question. What is interesting about many of these systems is how they have stood the test of time. The modern versions in cities such as London and Hamburg (the first such system constructed in Germany) have changed remarkably little in either form or principle. For our purposes, and placing ourselves in the position of a city official in the nineteenth century, they worked remarkably well. I give the nineteenth-century sewerage system, and especially that constructed in London, the maximum score of 4.

Novelty: Not only were sewerage systems of the type we have analysed a vast improvement on the more general problem of waste removal and sanitation, but they reveal massive shortcomings in the solutions that went before. Prior to the implementation of Bazalgette's London system, human waste—*night waste* as it was sometimes known—was frequently simply dumped in the cellars of poor dwellings, or emptied into the existing street gutters that directed water into the Thames. As a result, contamination was rife, diseases such as Typhoid were widespread, and the smell was often unbearable. A system that removed all of this waste quickly and efficiently to a location well away from the city, and which did not pollute drinking water, seems ridiculously obvious in hindsight. At the time, however, it must have been hailed as a great, albeit incremental, leap forward. I give these sewerage systems a score of 3 out of 4 for novelty.

Elegance: Bazalgette's London sewerage system, was, like many Victorian era engineering works (think of the beautiful railway stations that still grace many European cities) almost a work art! Indeed, The Observer newspaper, in April 1861 described it as "…the most *beautiful* and *wonderful* work of modern times…" This perhaps reflects the fact that the science and method of engineering was, by the Romantic Period, quite well understood. The systems created, from railways lines, to tunnels, bridges and ships were not always without mishap, but the engineers of this period understood how to define a problem, how to consider alternatives and how to implement those alternatives aided by the use of models, diagrams and rigorous analysis. From the selection of the main arteries of the London sewerage system, to the creation of the main tunnels, the design of pumping stations and the placement of the outfall, the London sewerage system is a true "system" in the modern sense. Its components fit together in a logical fashion, and a layperson can readily understand the way in which waste is collected and progressively directed to its point of discharge. For this reason, Bazalgette's system scores an impressive 3.5 out of 4 for elegance.

Genesis: Sewerage systems are an interesting example of the interplay of "new solution to an existing problem" (i.e. incremental innovation) and "new solution to a new problem" (i.e. radical innovation). Since humans first began living in groups, there has been some sort of need to deal with our bodily waste. Even prehistoric cultures understood that this was a problem of health and a problem simply of making a settlement pleasant to live in. To the extent that nineteenth-century sewerage systems modified the paradigm of human waste treatment—some attempt to remove it completely, rather than just dumping it—I give these systems a score of 2 for genesis.

Total: The sewerage system developed for London in the 1850s scores a total of 12.5 out of 16 on our scale of creativity. In my simple delineation of four bands for innovation (low, medium, high and very high), sewerage systems sit on the boundary between *high* (9–12) and *very high* (13–16). Its only real weakness lies in genesis: while it is strong in some aspects of novelty, it remains, at its heart, a new solution to an old problem. What would it take to change the paradigm of the treatment of human waste? Even today, we rely on systems that are little more than improvements in Bazalgette's system. A paradigm shift—a new solution to a new problem—might, for example involve transforming human waste, at the point of production, into a useful, reusable resource (e.g. fuel).

The Electric Lightbulb (1879)

The most certain way to succeed is always to try just one more time – Thomas Edison, American Inventor (1847–1931)

The electric lightbulb is really a very simple energy-handling system. Electrical energy is readily converted into heat and light, much the same as burning wood converts chemical energy to heat and light. The advantage of the lightbulb is that it does this process, if not more efficiently, then at least with potentially fewer pollutants. Even though we are now seeing the negative impact of humankind's voracious appetite for energy, it's not hard to imagine how much worse our climate system might be if every household burned coal or wood for heat and light. For decades, much of our global electricity supply has depended on coal-fired power *stations*—certainly not clean, but at least their efficiency could be optimised, unlike coal or wood burning in individual homes. There can be no doubt that the electric lightbulb, and domestic electricity more generally, has had a far-reaching effect on the world, in little over 100 years. This is another invention that satisfies needs at every level of Maslow's hierarchy, and yet our relationship with the lightbulb, and with electricity, is changing rapidly, as the effects of climate change make their presence felt.

What Was Invented?

The story of the invention of the electric light is not only one of the inventions, but also one of the competition and intrigue. Unlike the situation in many earlier time periods, by the time of the late Romantic Period, invention was not only a way to meet the needs of society, but also a highly commercialised enterprise—it was a way to make money. It seems that Thomas Edison was acutely aware of this, and very conscious of whom he was competing against, in the race to meet some very clear needs in the US society.

As populations grew, and as wealth swelled in an increasingly industrialised and urbanised society, demands for modern amenities such as artificial lighting and other public utilities grew. Gas- and oil-based lighting systems had been available since the late 1700s, and were first applied on a citywide scale in the early 1800s. Cities such as London and Paris began to be lit before 1820. By the late 1850s, gas lighting was widespread in England and is credited both with a rise in literacy, as people could read more easily during long winter nights, but also with negative social impacts resulting from the fact that factories could operate around the clock. Indeed, gas lighting systems are sometimes credited with stimulating the Second Industrial Revolution.[2]

By the late 1800s, however, electricity was maturing as a power source. This was not without problems, however, and the first of these was not an issue of the electricity itself, or its production and transmission. Nor was it a problem of converting the electricity into light. As early as 1800, Alessandro Volta had demonstrated that a wire could be made to glow by the application of an electric current. Instead, the chief problem faced in the late 1870s was how to make a lightbulb that would last for a decent length of time. This is the problem to which Edison first applied himself, it should be noted, as a very clear stepping stone to the more important and lucrative problem of creating an electrical power utility for America's growing cities (Fig. 3).

Much has been written about the related battle to establish the dominant form of electrical supply—alternating current versus direct current—and Edison ended up on the losing side (direct current), but only after some vicious skirmishes between rival companies. Probably, the least edifying was Edison's attempt to have the first electric chair made using his rival's AC system, as a means of demonstrating its inherent danger! However, all of that is incidental to our current purpose, and that is to examine Edison's invention of the longer lasting, and commercially practical, lightbulb.

Why Was the Electric Lightbulb Invented?

As much as it seems like we should identify the problem as *how to make electric light*, I have already noted that Edison was aiming, at least in part, at a larger and more lucrative problem. This suggests that we might have to consider leaning towards a problem like how to promote the use of electricity? Given his track record, it's hard to believe that Edison did not have a foot in both the altruistic camp—build a better

---

[2]We have already explored the First Industrial Revolution, characterised by the impact of steam power and mechanisation. The Second Industrial Revolution is usually associated with the impact of electricity as the source or power, however, the hypothesis here is that it was stimulated by changes arising first from the impact of widespread gas lighting.

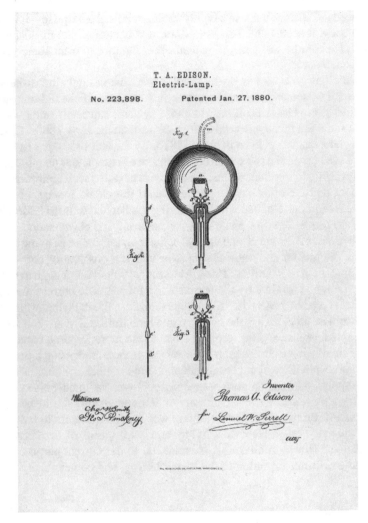

Fig. 3 Lightbulb patent application, 1880 (*Credit* Wikimedia, Thomas Edison)

lightbulb, because that's a good thing to do—as well as the capitalistic camp (how to monetise a good idea). While there are differing schools of thought on the interplay between capitalism and common good, it's hard to ignore the evidence that an awful lot of inventions have been stimulated by both a desire to tackle a problem and make money doing so.

How Creative Was the Electric Lightbulb?

Relevance and Effectiveness: Edison's work gives us very clear insights into the systematic development of inventions. His lightbulb was a necessary step in the replacement of gas as a means of domestic and commercial lighting. Edison understood, very clearly, the limitation of existing electrical lights: short lifespans (often only minutes, or a few hours), a high manufacturing cost and the need for high operating currents (which created problems of supply). He, therefore, set out to develop a light that specifically addressed these issues. In a remarkably short period of time, Edison focused his efforts on carbon filaments—in other words, the wire in the lightbulb that glows when current passes through it—and quickly succeeded in increasing the lifespan of the bulb from about 12 h up to 1200 h, with a carbonised bamboo filament. With this sort of lifespan, and using a readily available material, lightbulbs were both technically, and commercially, feasible. Edison's lightbulb, therefore, gets the maximum score of 4 out of 4 for relevance and effectiveness.

Novelty: Edison's electric lightbulb is an excellent example of a solution that highlights the weaknesses of the existing artefacts. It lasted hundreds of times longer than the examples, it was far cheaper to produce, and it was designed to operate with low currents. Not only that, but it was a case study in how the key weaknesses of previous attempts could be systematically addressed. At the same time, it was not a radically new approach. The filament concepts already existed, and simply needed to be improved. Therefore, while possessing many strengths in this category, Edison's lightbulb did not shift a prevailing paradigm, and its surprise value was limited to particular internal details—for example, who would have thought that a carbonised bamboo filament would be the answer? It was still a filament, just one of an unusual material. Overall, therefore, Edison's lightbulb scores a healthy, if not brilliant, 3 out of 4 for novelty.

Elegance: This criterion looks at the quality of the invention. Is it well made, and pleasing in appearance? This may seem rather unconnected to innovation and the solution of problems, but it is remarkable how strong a connection there is between looking nice, and working nice. Engineers know this, and in fact, there is more than one piece of engineering wisdom that reminds us of this. Wernher von Braun—the German V-rocket scientist in World War 2, turned NASA Saturn V designer—is reputed to have said that "the eye is a fine architect, believe it". What he meant was that engineers, and indeed many non-engineers, instinctively recognise good engineering solutions when they see them. I often show people two photographs of bridges. One is made from tree trunks haphazardly laid across a mountain chasm, while the other is a picture of a modern steel and concrete road bridge. I then ask people which one they would prefer to drive over. The answer is always the steel and concrete bridge. The reason, I am certain, is because you don't need to be a civil engineer to see that the tree trunk bridge doesn't look very good. It looks dangerous, unstable and unlikely to support a vehicle, while the steel and concrete bridge looks as though it will still be standing 200 years from now.[3] We seem to have an innate ability to recognise good solutions. What we seem to be evaluating, even without realising it, is that artefacts that look well executed, nicely proportioned and complete, *usually* are more effective. So the more elegant a solution is the more effective it seems to be. In the case of Edison's lightbulb, it looks pretty good. People seeing it for the first time in the early 1880s must have had reasonably high confidence based on its elegance. Perhaps, it was a little fragile, but apart from that, it looked like a pretty good solution, and it turns out, it functioned pretty well. I give it 3 out of 4 for elegance.

Genesis: I have already hinted at the one weakness of Edison's lightbulb with respect to general qualities of surprisingness and in how it fitted the prevailing paradigm. These same concerns weaken the potential score for genesis, which rewards solutions that really begin to not only break a paradigm, but also define new problems and solutions that emerge from the current solution. While the lightbulb goes some way down this path—for example, it suggests that there might be further opportunities, beyond Edison's design, if a similar approach is continued (i.e. there might be other, even better materials for filaments or other ways of prolonging the life of existing filaments)—it still falls short when it comes to new concepts and new paradigms. In simple terms, it was backward-looking, seeking to improve what came before, rather than forward-looking, seeking to start a fresh, new approach. For these reasons, I have given the lightbulb 2.5 out of 4 in the category genesis.

Total: Edison's incandescent lightbulb earns a score of 12.5 out of 16 for creativity. This places it in what is becoming a special *demilitarised zone* for famous inventions, sitting on the boundary between high and very high creativity. Two things stand out as slight weaknesses, and by now, they are familiar to us. First, this fledgling commercial-scale lightbulb is very incremental in nature. I suspect that even Edison would have admitted as much. It was a systematic, deliberate attempt to fix the weaknesses of the existing incandescent paradigm, in order to make the technology commercially viable. The weakness in our creativity scale is, therefore, a relative lack of true surprise value and paradigm-breaking qualities. This tends to pull down the scores both for novelty and genesis. A second slight weakness lies in the combination of effectiveness and elegance—i.e. functionality. As good as the lightbulbs were, they still had room for refinement, as modern bulbs demonstrate. While it is unfair to judge them with the knowledge we now have, I can imagine people at the time still lamenting their relatively limited lifespan, or a fragility in the design that detracted a little from their usefulness as a solution. Perhaps if Edison had discovered, and then applied, a completely new principle for electric lights, or if his bulbs had been even longer lasting and more robust, we would be talking about an invention in the upper reaches of the creative solution scale.

---

[3]The tragedy of the bridge collapse in Genoa, Italy, in August 2018, shows that visual appeal, however, is not sufficient on its own.

The Modern Age (1880–1950): Expansion and Conflict

The Modern Age, defined largely as the first half of the twentieth century, saw some of the best, and the worst, consequences of humankind's inventive ability. Few families even today are untouched, in some way, by the effects of the two great, global conflicts of this epoch, and few were spared the economic effects of the Great Depression. At the same time, from these catastrophes emerged a world poised to reap the greatest benefits of creativity—massive improvements in health, enormous improvements to education, trade and welfare and the promise of an end to global conflicts with the invention of a mechanism for resolving disagreement and conflict on a global scale. An interesting question that often arises is—would the innovation of this period have been the same without the First and Second World Wars? There is some compelling evidence that wars stimulate technological creativity, perhaps due to a greater appetite for risk and fewer organisational barriers. Were the advances of the Modern Age accelerated because of these conflicts, or would they have proceeded just the same?

The first invention that we consider in this era is entirely unrelated to conflict, and it is the steel-girder skyscraper. This advance changed the kinds of structures that could be built, and saw a shift in cities to building *up*, further concentrating populations in small areas. The second innovation we consider is another—like the printing press—often cited for its profound impact, and is the first powered aircraft, the Wright Flyer. Although this was not my intention, it's curious, in the light of the discussion about conflict, that skyscrapers and planes would herald a new era of conflict almost 100 years after their invention. The final innovation that we will consider in the Modern Age is another that has a link to conflict, but in a good way. Antibiotics—penicillin specifically—was discovered by Alexander Fleming in 1928, but it took the stimulus of the Second World War to drive the invention of large-scale, useable antibiotic drugs that have been credited with saving millions of lives.

© Springer Nature Singapore Pte Ltd. 2019
D. H. Cropley, *Homo Problematis Solvendis—Problem-solving Man*,
https://doi.org/10.1007/978-981-13-3101-5_10

Steel-Girder Skyscrapers (1885)

New York is vertical: all skyscrapers—Tony Scott, British Film Director (1944–2012)

Previously, in our discussion of the construction cranes of ancient Greece, we noted that tall buildings can be thought of as a kind of battery—an energy-handing (storage) system. Raising a mass against the force of gravity is a process of giving that mass so-called gravitational potential energy. In the case of energy storage, our goal is to release that energy at some later point, for example to create heat or light. In the case of the skyscraper, the trick is locking in that potential energy, and preventing it from converting back to kinetic energy (by falling). William Jenney (1832–1907) simplified the process of storing gravitational potential energy in buildings, first by reducing the amount of energy that had to be supplied (by reducing the mass of what had to be lifted) and second, by finding a more effective way to keep it locked in as potential energy. Once this was achieved, useful things, like living spaces or offices, could be created in the resultant structure. This satisfies a range of humankind's needs, from warmth and shelter, through safety and security, and, in structures like the Sydney Opera House (principally, a national icon), can even satisfy the need for esteem and self-fulfilment.

What Was Invented?

Prior to the advent of modern, multistorey commercial and residential buildings—skyscrapers, in other words—in the late nineteenth century, tall buildings were more aesthetic than functional in nature. The great European cathedrals of the Middle Ages (e.g. Chartres in France, built in the late twelfth century) were often impressively tall, but constructed principally to glorify God. Not only that, but their size was only possible through the construction of a kind of exoskeleton—an external framework of large pillars and buttresses—necessary to support the large and heavy structure. It is noteworthy that these earlier buildings, though often tall, did not generally support much in the way of living or working spaces at great heights, because the additional mass could not easily be supported at these heights.

The ancient Romans constructed multistorey apartment buildings. These were typically up to about 6 or 7 stories in height, and were constructed from timber, mud brick and concrete. They typically lacked amenities such as running water or heating in the upper floors, and of course had no elevators, so that the cheapest apartments were usually those at the top. They also suffered from the significant disadvantage that they were at risk from fires, and prone to collapse!

This all changed with the invention of the steel-girder skyscraper, first developed and built in Chicago by William Jenney in 1884–85. The breakthrough was the use of an endoskeleton—an *internal* supporting framework—made of steel beams. These columns and girders formed the basic structure of the building onto which so-called curtain walls were fixed. Unlike previous designs, the building's walls were no longer the principle weight-bearing structure, and therefore, could be made from much lighter materials, including glass. The reduction in weight, as well as the strength of the steel endoskeleton, meant that buildings could be much higher, with the proper use made of all levels in the building.

Jenney's first design was a 10-storey office building in Chicago—the Home Insurance Building (see Fig. 1)—and is estimated to have weighed one-third of an equivalent masonry building. As well as the basic steel frame skeleton, the building also used reinforced concrete in its design. While neither of these makes a building completely fireproof, they do significantly reduce the risks from a large fire.

Why Was the Steel-Girder Skyscraper Invented?

The steel-girder Skyscraper is another example of proximate and distal problems. Jenney's basic problem was how to build higher structures, but to solve this we can argue he had to solve the problem of how to efficiently store potential energy in buildings. This, in turn, could be said to be a problem of how to create a lightweight, internal skeleton for a building. These are all interrelated and different expressions of the same underlying issue. We can usefully state these all as the fundamental problem of *how to build upwards*?

How Creative Was the Steel-Girder Skyscraper?

Relevance and Effectiveness: The steel-girder skyscraper—both the first example built by Jenney, and many subsequent examples—demonstrates a high degree of relevance and effectiveness. These buildings immediately demonstrated their ability to solve the problem of building *upwards* with a high degree of safety and reliability. They represented a pinnacle and integration of knowledge and technology of the period, and showed what was possible. This resulted in the rapid development of the technology, with buildings rising ever higher. For these reasons, it is hard to give a score of anything less than the maximum of 4.

Novelty: The steel endoskeleton concept immediately drew attention to the weaknesses and deficiencies of previous construction methods, and shows not so much how other buildings could be improved, but how they are obsolete. The new approach embodied in this construction method also suggests how the technology can be improved further, applying some of the concepts to build higher, and lighter buildings. These qualities give the steel-girder skyscraper a score of 3.5 for novelty.

Elegance: I have given the steel-girder skyscraper a score of 4 for elegance. Simply looking at photographs of Jenney's original design, which was demolished in 1931, we can see a building that is skilfully executed, pleasing to the eye, well proportioned, and complete. It looks like a solid, well-constructed building, confirming a common

Fig. 1 Exterior of the Home Insurance Building (*Credit* Wikimedia, US Library of Congress)

engineering rule of thumb—namely, that a good solution usually looks like a good solution.

Genesis: The steel-girder skyscraper has strong elements of this criterion, reflecting the fact that it goes some way towards changing our understanding of the problem of building upwards. It embodies a novel basis for further work on this same problem—other strong, yet relatively lightweight materials, other materials (such as glass) to form the external skin of the building. It also prompted the application of similar concepts in similar ways—the Eiffel Tower, opened in 1889, is essentially a steel-girder skyscraper with no external skin! The skyscraper, furthermore, draws attention to new problems inherent in this new design. A good example has been the problem of pumping cement up to the top of very high modern skyscrapers during their construction. Traditionally, concrete would be lifted in large buckets by crane, however, modern advances now make it possible to pump concrete to heights in excessive of 700 m (e.g. in the Burj Dubai Tower in the United Arab Emirates). In short, the skyscraper changed much of how we understand the problem of building tall, very functional structures, and therefore scores 3 out of 4 for genesis.

Total: With a total score of 14.5 out of 16, the steel-girder skyscraper literally, and figuratively, stands taller than most of our innovations so far. This score, as you would expect, places it in the middle of the *very high* range for creativity. The steel-girder skyscraper may well be something of a surprise on our scale of creativity, however, it may well be the near-perfect example of *invention*. It was not only immediately highly *functional*—i.e. both effective and elegant—and not only highly incrementally novel, but it opened up many new perspectives on the creation of buildings. Although a new solution to an existing problem, the true test of genesis is perhaps the extent to which it rendered previous and competing approaches obsolete. The steel-girder skyscraper seems to have done this to a very high degree.

The Wright Flyer (1903)

If birds can glide for long periods of time, then… why can't I?—Orville Wright, American Aviation Pioneer (1871–1948)

It is tempting to see the Wright Brothers' achievement in December 1903—the first powered, sustained, controlled and heavier-than-air flight (an energy-handling system)—as a radical innovation, and one that has been enormously influential in modern history. However, the history of flight more generally makes it clear that the Wright Flyer was simply the latest in a long succession of attempts by humans to take to the air. From the rather naïve and hopeful tower jumping of the Middle Ages, to kites, hot air balloons and gliders, mankind has had a fascination with flight. Has this fascination been driven by basic physiological needs? Probably not, although, like many of our inventions, it has contributed to our never-ending quest to satisfy our need for food (e.g. through transport, or even, through activities such as crop spraying). The underlying motivation, in fact, may be much higher in Maslow's pyramid. From the earliest days of Daedalus and Icarus in ancient Greece, and their attempts to soar like birds, has the real driving need behind humankind's quest to fly been simply one of self-actualisation? Has it been our desire simply to achieve our full potential—a kind of transcendental creativity—that has motivated humans to solve the problem of flying?

What Was Invented?

The Wright Flyer was the first airplane in the modern sense of a manned, powered, flying machine. It achieves flight by solving two problems. First is the question of how to create lift. This is necessary to generate the force that keeps an otherwise heavier-than-air device firmly on the ground. This also highlights an important distinction from lighter-than-air objects that simply *float*, like balloons. The second problem that the Wright Flyer had to solve was the question of how to generate thrust. This is the horizontal force that pushed the aircraft through the air. The Wright Brothers solved the first problem with a lift-generating wing, although they did not invent this, and solved the second problem with a lightweight petrol engine and propellers.

The Wright Flyer itself was constructed from wood taken from the giant Spruce tree. It employed a so-called bicanard design, with two small, horizontal control surfaces that determine the pitch of the aircraft (whether the nose is pointing up or down) in front of the pilot (see Fig. 2), and not at the rear as is usually the case today. A pair of vertical control surfaces that determine the yaw of the aircraft (turning the nose left or right) was at the rear. Banking and rolling were controlled not by flaps on the wings, as is normal nowadays, but by warping (i.e. twisting) the entire wing structure through a system of wires and pulleys connected to the pilot's hips. The wings were covered in unbleached muslin fabric (a light cotton weave).

The entire structure weighed approximately 340 kg including the weight of the pilot, had a wingspan of 12.3 m and a length of 6.3 m. The petrol engine was a 4-cylinder, water-cooled engine generating about 12 horsepower (about 8.95 kW)—a typical, modern small car has an engine of about 80–90 kW.

To assist them in their efforts to achieve the first powered flight, the Wright brothers used a combination of an inclined launch ramp (to generate some gravity-assisted speed) or headwinds. Their first attempt, on 14 December 1903 and piloted by Wilbur, used the inclined ramp, but he pulled the nose of the aircraft up too sharply, stalled and had a minor crash, resulting in some damage. On 17 December 1903, having made their repairs, Orville took the controls with a flat launch into a headwind of about 32 km/h, achieving a controlled flight lasting about 12 s and covering 37 m (so his average speed over the ground was just under 11 km/h—a reasonable jogging speed). The brothers continued to take turns on the same day, making a total of four flights of increasing distance, ending the day with a 59 s flight of 260 m (about 15.5 km/h—more of a serious running speed). This particular aircraft was subsequently damaged by a wind gust, and never flew again—however, history had been made.

Fig. 2 First successful flight of the Wright Flyer, 1903 (*Credit* Wikimedia, US Library of Congress)

Why Was the Wright Flyer Invented?

As indispensable as powered flight has become in just over 100 years, it seems hard to make a case that the Wright Brothers were setting out to solve anything other than the highest level of Maslow's hierarchy of needs. They could not have anticipated just how useful aircraft would become in addressing basic, and indeed psychological, needs, but their goal seems to have been little different from Greek mythology's Daedalus and Icarus, with the feathers and wax wings. The Wright Brothers were seeking to solve "the Flying Problem"—simply, how to allow humans to fly. It was, to some degree, the problem of "how to let Orville fly like a bird". We do know, however, that they understood a crucial sub-problem to be the issue of control. Humankind knew how to make wings that generated lift, and they realised that a means was required to generate thrust—the real trick, however, was how to control this new invention, and this is what the Wright Brothers did. The solved the problem of *how to control powered flight*.

How Creative Was the Wright Flyer?

Relevance and Effectiveness: I thought I had surprised myself when I could only award a score of 3.5/4 to the Julian calendar, and yet here, we have another case of an invention that scores less than the maximum for relevance and effectiveness. As important as the Wright Flyer was to the development of aviation and as much as it fitted the basic constraints of the problem—achieving powered, heavier-than-air and controlled flight—the particular model in question did this, but only just. Its performance, if we are being strict, was not great, and that pulls down its score in this category. This does not mean that it was necessarily deficient. Rather, I think it reflects the fact that the Wright Flyer was very much an experimental vehicle. It was never going to be perfect, and therefore, perhaps somewhat paradoxically, it can probably never be rated as highly effective, even if it was relevant to the problem. It's a little like Roger Bannister breaking the 4-min mile barrier. His time of 3:59.4 was never going to remain the best ever, but it showed what could be done, and like the Wright Flyer, was a stepping stone to greater achievements. For this reason, the Wright Flyer gets a score of 3 out of 4.

Novelty: The Wright Flyer was important in helping to define what aviation pioneers like Orville and Wilbur Wright were trying to achieve. It drew attention to weaknesses in other aircraft of the day, and highlighted some of the component problems that needed to be solved if the craft was to fly—for example, the weight and power requirements of the engine. The Wright Flyer also helped the brothers to collect information about their experiments. For example, they placed the so-called horizontal stabiliser in front of the pilot (and not behind as is common in modern aircraft) specifically so that the pilot could see the position of this control surface without having to turn around and look behind him (remembering that they had no instruments on this aircraft). For these reasons, there is much that is new in the design that contributes to the ongoing development of aircraft, and I give the Wright Flyer a score of 3 out of 4.

Elegance: As crude as this invention may look to our modern eye, there is still much to admire about how it was executed by the Wright brothers. Both drew on their experience as bicycle makers to execute a skilful and well-constructed aircraft. Although we are used to very different designs nowadays, the design of the Wright Flyer—given its highly experimental nature—is nevertheless complete and well proportioned. The brothers invested a lot of effort in this innovation, and its quality is reflected in the fact that it was successful in achieving powered flight. I give the Wright Flyer a 3 out of 4 for elegance. Even today, it looks like a good solution to the problem.

Genesis: As with many of our innovations, genesis is probably the hardest quality to achieve well. This is true of the Wright Flyer. It changed elements of the understanding of the problem in question, but as with many of our inventions, it was far more incremental than it was radical. It helped identify areas for improvement, and it drew attention to weaknesses of previous designs. It established a new benchmark for powered flight. At the same time, it did not radically change the underlying approach to aviation, nor did it open up new conceptualisations of other problems, so that I score it 2.5 out of 4 for this category.

Total: Overall, the Wright Flyer achieves a score of 11.5/16 on our scale of creativity. This places it comfortably in the "high" range, with some elements holding it back from a higher score—notably its experimental, developmental nature. I have to keep reminding myself that 11.5 does not mean an invention is *uncreative*, and this is the case for the Wright Flyer. However, when we look at it in the grand scheme of things, and even as it must have appeared in 1903, it was not a mind-blower. There were many other inventors very active at the same time, all seeking to solve the flying problem, and I can imagine that the reaction was more a confirmation of expectations than disbelief. That takes nothing away from what was still a momentous achievement.

Antibiotics (1940s)

One sometimes finds what one is not looking for— Alexander Fleming, Scottish Physician (1881–1955)

The last invention of this era—the Modern Age—is a material handling system. Penicillin, the antibiotic drug derived first from the mould strain *Penicillium notatum*, and later from *Penicillium chrysogenum*, kills bacteria cells by interfering with their ability to build new cell walls and to reproduce. Unlike modern gene therapies, which derive their therapeutic effect from the information stored in DNA (we will discuss these in our final epoch), and unlike the use of the high energy X-rays to treat cancer, this medical innovation employs one material agent (mould spores) to kill another material agent (bacteria cells). The direct need that all of these treatments satisfy is a basic, physiological one. However, a secondary effect of any successful medical treatment is that higher order needs are facilitated once a person is fit and well.

What Was Invented?

I have often told the story of how Alexander Fleming discovered penicillin and will tell it again here. However, if you've been following my thought process carefully, you will recall that *discoveries* are outside of the scope of our analysis. In fact, if we were to rate the creativity of Fleming's discovery of penicillin, it would fare rather poorly. The problem he was trying to solve related to the antibacterial properties of nasal mucus. So penicillin would have to score poorly for relevance and effectiveness. It was, however, highly surprising, and would score well for novelty. Unfortunately, creativity requires the minimum condition of effectiveness *and* novelty, so Fleming's discovery violates an important precondition. Aside from those two criteria, it would have to be rated as an inelegant solution, even to the real problem of killing infections in people. The fact is, Fleming's *P. notatum* was difficult to produce in quantities sufficient for clinical purposes, and could not simply be scraped off a Petri dish, mixed with some water, and injected into humans. It was an incomplete, obscure solution, and to a casual observer, and possibly even many scientists, it would not have been obvious that what was taking place in Fleming's Petri dishes was a miraculous medical breakthrough (see Fig. 3).

Fig. 3 Sample of penicillium mould presented by Alexander Fleming to Douglas Macleod, 1935 (*Credit* The Science Museum, Christie's South Kensington)

Indeed, we know that it was not obvious, because some 50 years before Fleming, a German scientist, Eugene Semmer, effectively discovered penicillin, but ignored the antibacterial fungus as an inconvenient obstacle to his research on trying to cure horses of infections. Finally, and on a more positive note, we would probably give Fleming's discovery a high score for genesis precisely because it changed a paradigm, and opened up an entirely new perspective. Perhaps, this illustrates a key difference between inventions and discoveries—the former is directed to the solution of a problem, which demands not just novelty and genesis, but also effectiveness and elegance; for the latter, novelty and genesis are the key, and effectiveness and elegance come later.

So what is this section about, and who is/are the heroes of our story? If Fleming discovered penicillin, then it was two other scientists—Howard Florey (1898–1968) and Ernst Chain (1906–1979)—who *invented* antibiotics and made these life-saving drugs available on an industrial scale. To be fair, Fleming was one of the heroes too, and the three men shared the 1945 Nobel Prize for Medicine.

Our invention, therefore, is the antibiotic *drug*, based on the strain *P. chrysogenum*, which Florey and Chain found to produce a higher rate of penicillin, and was therefore more suitable for development into a viable, large-scale medicine. During their search for a more productive strain, it was a fungus growing on a cantaloupe in a grocery store in the US that turned out to be the solution. Florey and Chain then irradiated samples of this fungus with X-rays and UV light, causing a mutation that yielded an even higher rate of penicillin—more than 1000 times the rate of Fleming's original *P. notatum*—and the basis was set for large-scale manufacture of antibiotics.

Why Were Antibiotics Invented?

Building on Fleming's serendipitous discovery, it was immediately apparent that penicillin had important medical applications. Humans had, for millennia, been at the mercy of bacterial infections, so that it seems fair to say that the need—how to cure bacterial infections—was a problem waiting for a solution. As soon as a promising solution emerged, in the form of penicillin, the race was on to develop this from a promising idea to a practical treatment. The clear, driving problem exercising the minds of Florey and Chain was undeniably *how to cure infection*.

How Creative Were Antibiotics?

Relevance and Effectiveness: As someone who was hospitalised seven times in 1 year—at the age of 8—with pneumonia, I can attest to the effectiveness and importance of antibiotics. I can still remember the injections, given at midday and midnight! It was terrifying to be woken in the middle of the night, and stabbed in the leg with a needle, but it was necessary! There can be little debate for this category. Some estimates have put the total number of lives saved, since the invention of antibiotics, at 200 million. In WW2, it is estimated that antibiotics saved the lives of 12–15% of wounded Allied soldiers. This invention solved the problem of curing infections, and antibiotics get the maximum score of 4 out of 4.

One side note here. As always, we put ourselves, as far as we can, in the time and place of the invention in question. In the 1940s, antibiotics were a true miracle drug. In fact, they remain so, but there is a problem. Many bacteria have evolved to be resistant to antibiotics, and this is a real and serious medical problem. Without the discovery of new antibiotics, we face the frightening possibility that people will, once again, succumb to infections that have not been regarded as problematic for 60 or 70 years. What this means in creativity terms is that the effectiveness of antibiotics is declining. In fact, this seems to bear out an important point I made in an earlier chapter. The effectiveness of antibiotics is not changing for no reason. It is declining because, in effect, the novelty of antibiotics has declined, and we know that declining novelty causes a decline in effectiveness. In a very real sense, bacteria have now *known about* antibiotics for close to 100 years. Like any entity engaged in a competitive, creative process, whether a technology company or terrorists on a hijacked airplane, the actors in this process rarely sit still. Companies respond to their rivals with their own new products; passengers respond to terrorists by fighting back; and apparently, bacteria respond to our creative medicines, by evolving drug-resistant strains. If we judge the creativity of antibiotics with a twenty-first century eye, we might find a worrying decline in effectiveness, which knocks a couple of points off the score. It's time to launch the equivalent of the next iPhone of antibiotics, and quick!

Novelty: Antibiotics represented a considerable departure from the existing approaches to treating bacterial infections. Prior to the identification of various strains of penicillin and their practical application as antibiotics, the doctors used Sulpha drugs (invented in the 1930s, and which merely inhibit the growth of bacteria, but don't kill them), or might try a blood transfusion, or even herbal remedies. These were of limited effectiveness and the underpinning concept of antibiotics was fundamentally different—and new. Antibiotics showed how the existing treatments could be improved, and highlighted the shortcomings of previous treatments. They indicated both a radically new approach, and quickly stimulated ideas for new applications of the drug—treating many different kinds of infection, and developing new variants based on the same underlying principles. For these reasons, antibiotics score well for novelty, with 3.5 out of 4.

Elegance: I have given Florey and Chain's invention a score of 3 out of 4 for this criterion. It is clear that, in broad terms, the original antibiotic drug was a well-executed solution to a well-defined problem. The reason I have not given it the maximum score for elegance is twofold. First, I have to ask myself if I am assessing the drug itself, or the means for producing the drug. If it is the former then it is, in some ways, more a lucky solution than anything else. If it is the latter, then I believe it is less *elegant solution*, and more brute force. The fact is, a usable drug was possible, based on Fleming's original discovery. Florey and Chain simply (I say this with no disrespect intended) refined and scaled up the process to mass produce the drug. I feel it is hard to describe it as a beautiful or aesthetically pleasing solution. It simply required the know-how to generate large quantities of the drug. 3 out of 4 is not a bad score, it just recognises that there are aspects of the mass production of antibiotics that would be progressively refined and improved, once the immediate imperative of saving the lives of soldiers had passed.

Genesis: Florey and Chain's innovation scores fairly well in terms of genesis with 3 out of 4. Not only are antibiotics a novel solution, but their development stimulated decades of further research and development, with scientists finding many more forms of antibiotics suitable for particular kinds of infections, and ever more efficient in treating infections. The underpinning concept of antibiotics also contributed to new approaches to the treatment of other medical conditions. Rather than searching for what might be termed more passive, chemical methods for inhibiting or killing bacteria, antibiotics used a living organism—fungus—to actively fight and destroy another living organism. This, in turn, has contributed to new approaches to the treatment of other kinds of disease, and therefore, antibiotics score quite highly for genesis.

Total: Florey and Chain's development of antibiotics, vitally underpinned by Fleming's discovery, scores a very respectable 13.5 out of 16. This *very high* score for creativity is held back only by slight deficiencies—perhaps somewhat unusually—in elegance and genesis. As I write that, I feel as though I'm short-changing these inventors (and keep in mind that Howard Florey is a native of my hometown in Australia) but objectivity here is vital. Perhaps, we can argue that elegance was not vital to this innovation—it didn't need to be beautiful, it just needed to work. Antibiotics would be refined substantially in the decades following their first mass production, but our analysis always takes place at the time of invention. I also feel that a higher score for genesis could be justified, except that the origin remained a discovery. Florey and Chain were therefore working primarily incrementally, in response to Fleming's discovery, and this always limits the absolute genesis possible.

The Space Age (1950–1981): The Science of Creativity

The penultimate period that we will tackle is the *Space Age*. For me, born in 1967, this still looms large as an age of exciting, grand achievements: landing humans on the Moon; the Space Shuttle; spacecraft sent to the outer reaches of the solar system. I still find it thrilling to read about the Apollo program, and the ingenuity with which NASA tackled multiple technological challenges using rudimentary digital computers and other equipment that seems almost laughably simplistic to a twenty-first-century eye. The computer on board the lunar landing module of Apollo 11 had 2k of memory and ran at a speed of 1 MHz (compared to a typical modern smartphone with a clock speed that is more than 1000 times faster, and usually has at least two or three million times as much memory). I have a vague memory of watching the 1969 Apollo 11 moon landing on a grainy black and white TV, and throughout my childhood, the epithet "Space Age" was applied to anything that people wanted to show was super-modern.

As you probably realise by now, I'm about to explain that you may be surprised to see that not everything in this chapter is, therefore, about space, rockets and moon landings. Indeed, the one space-related invention that I am going to discuss is not even American, but comes from the Soviet Union, in the form of the world's first artificial satellite—*Sputnik*.

Because the Space Age was not just about extra-terrestrial inventions, our first invention will focus on *nuclear fission*. Not *the Bomb*, however, but controlled nuclear fission—the splitting of atoms—to produce usable energy. This need for energy is an ever-present one across humankind's history. Finally, we will close out this chapter by turning our gaze from the heavens to the microscope, and peer inside our own bodies to examine genetic engineering and specifically what is known as *antisense (gene) therapy*. Together these innovations represent something of the impressive extent of our problem-solving ability, reaching from the tiniest molecules in our bodies, and invisible atoms, right up to the vastness of outer space.

© Springer Nature Singapore Pte Ltd. 2019
D. H. Cropley, *Homo Problematis Solvendis—Problem-solving Man*,
https://doi.org/10.1007/978-981-13-3101-5_11

The Nuclear Power Plant (1951)

I am become death, the destroyer of worlds – J. Robert Oppenheimer, American Theoretical
Physicist (1904-1967)

A nuclear power plant is perhaps the epitome of an energy-handling system. Enormous amounts of energy are liberated by subatomic processes, and the resulting heat energy is capable of powering almost everything that we take for granted in twenty-first-century societies. The needs that are satisfied by such an invention range from the most basic and essential physiological needs—food, water, warmth, shelter—right up to our loftiest needs for self-actualisation. At every level, the availability of plentiful, reliable, clean and affordable power—usually in the form of electricity—is essential to satisfying our needs. Although it is true, we have alternative ways of addressing all of these needs, nuclear power is possibly the most versatile. It can provide power a thousand feet under the ocean, and it can provide power in the farthest reaches of the solar system. It doesn't need oxygen, and is independent of the wind, the tides and sun. Of course, it has a certain image problem, but nuclear power may yet be a solution that we cannot ignore.

What Was Invented?

In December 1951, the world's first nuclear reactor designed to produce electrical energy became operational. The Experimental Breeder Reactor 1 (or EBR1), therefore, became the world's first nuclear power plant.

Nuclear *anything* is often a politically charged topic, and it's easy to understand why. The concept of the nuclear fission of heavy elements—that is, splitting atoms of elements such as uranium, to release large amounts of energy—was discovered in Berlin in 1938, but sat on top of several decades of research beginning with Ernest Rutherford's 1911 model of the atom. Because of the geopolitical state of the world in 1938—the rise of Nazism in Germany and a looming global conflict—nuclear fission quickly took on immense significance because of its potential as a weapon of mass destruction. Indeed, in August 1939, only weeks before the outbreak of World War 2, physicist Leó Szilárd wrote a famous letter, co-signed by Albert Einstein, to US President Franklin Roosevelt, warning him of the danger posed by this technology, and urging him to start a nuclear program in the United States. This led directly to the establishment of the Manhattan Project in 1942, which resulted in the development of the first nuclear bombs that were dropped on Japan. With the end of WW2, attention also turned to more peaceful applications of nuclear fission, resulting in the invention of the Experimental Breeder Reactor 1.

The underpinning concept of a nuclear fission power plant is as follows. A nuclear reaction—the splitting of atoms and the release of heat—which accelerates out of control in a nuclear bomb, is controlled in a reactor through a moderator. Typically, the moderator is a set of graphite rods that absorb some of the neutrons produced in the nuclear reaction, and allow the reaction to proceed more slowly. The graphite moderator is also the off-switch of the fission reaction, allowing the chain reaction to be shut down. In this way, a nuclear reactor can generate a large amount of heat from a very small amount of fissile material, but without the runaway reaction that causes a violent explosion.

The actual power generation in a nuclear power plant occurs by using the energy produced by the nuclear reaction to boil water, creating steam, and this steam is used to turn turbines that generate the electricity. This is exactly how electricity is generated in a coal-fired power station, except that the heating of the water is done by burning coal, instead of by the nuclear reaction. Of course, there are dangers associated with nuclear power if something goes wrong, and we saw this with the Chernobyl accident in 1986. However, properly operated and maintained, nuclear power production is, in many respects, far cleaner than burning coal or oil. We now know that coal and oil have caused enormous damage to the earth's climate system by injecting billions of tonnes of carbon dioxide into the atmosphere each year.

Whatever the perception of nuclear power, perhaps hampered by its association with nuclear weapons, this invention has been enormously influential over the past 60–70 years. Although safer, and greener, forms of energy production are now becoming available and viable, various forms of nuclear power production are likely to continue to play an important role for many years to come (Fig. 1).

Why Was the Nuclear Power Plant Invented?

Although the first application of nuclear fission was undeniably for the purpose of building a weapon, its potential as a source of power was also appreciated. In that context, the problem was the same as one that humankind has grappled with for millennia. In simple terms, that is the question of how to generate heat. Even in warm climates, humans have always needed heat, to cook food, and to provide illumination, so that it seems that we have been engaged in a constant quest to generate heat by one means or another. For thousands of years, as we know, the default source of heat was fire. However, fire has always had limitations. It may be hard to transport, or require large volumes of fuel, or occasionally run out of control and do damage. For that reason, many of humankind's inventions have involved either ways of controlling fire (and therefore, the heat and light it gives us) or replacing it altogether. For this reason, I am going to stick with the idea that the nuclear power plant solves the problem of how to generate (vast quantities) of heat.

How Creative Was the Nuclear Power Plant?

Relevance and Effectiveness: The first nuclear power plant, based on the Experimental Breeder Reactor 1, is an obvious candidate for the maximum score of 4 out of 4 in

Fig. 1 Shippingport power station reactor vessel (*Credit* Wikimedia)

this category. Alongside nuclear bombs of the same era, it was the acceptable, beneficial pinnacle of some 40 years of research in the field of nuclear physics, coupled with the translation of those theoretical and experimental concepts into a practical, problem-solving domain. It may be a reflection of my disciplinary background in both physics and engineering, but there is, to me, something almost heroic in how theoretical concepts that were unheard of only some 40 years earlier had matured to the point that almost unlimited quantities of energy could be produced from small quantities of raw material. More than many scientific disciplines, there is something almost unbelievable about the possibility of producing enough energy to power a city, simply by breaking apart tiny atoms. The EBR1 not only split atoms, and unleashed their energy, but did so in a safe, controlled and practical way.

Novelty: As effective as the EBR1 was, it has some weaknesses against the criterion novelty. To a certain degree, of course, it highlights shortcomings in other forms of power generation. Coal, for example but not because of what we now know in terms of climate change. Rather, nuclear power draws attention to weaknesses such as the need for a much more complex, and constant, supply chain for coal- or oil-fired power generation. A major weakness of coal-fired power stations is the infrastructure required (rail, for instance) just to keep the facility supplied with a constant source of coal. The EBR1, therefore, simultaneously shows what it looks like when a plant does not require this constant, high-volume supply of a raw material. On the other hand, nuclear power, and specifically the EBR1 doesn't score as highly as it might, because it introduces its own shortcomings, which are revealed when set against the existing solutions. This includes, for example the greater complexity of creating the fuel required for a nuclear reactor. As a result, the EBR1 has strengths when set against other solutions, but also some weaknesses, and this drags down its score for novelty. It is incrementally better, in some respects, but incrementally weaker in others. Therefore, I can only give it a score of 2.5 out of 4 for novelty.

Elegance: Not surprisingly perhaps, I make the case that there is a high degree of elegance in the world's first nuclear power plant. There is a neatness about the EBR1 not only in a literal, physical sense, but also in the broader system design. This first practical, commercial nuclear reactor is well executed. In fact, given the newness of the overall concept, it is almost remarkable that it worked so well. Perhaps this is a reflection of the fact that the first application of nuclear fission was in making a bomb. This emphasised the care and attention to detail needed to ensure that a nuclear power plant was safe. It's conceivable that if the first application of nuclear fission had been for power generation, physicists and engineers might have been less careful, possibly failing to appreciate the immense power that they were trying to control. In other words, the fission bombs made as the first real application of nuclear power demonstrated a worst-case scenario, and set a stringent benchmark for safety and care in the design of nuclear power plants. If elegance is an expression of good execution, careful and holistic design, then the EBR1 must be rated close to the maximum for this category. The only case I can make for a slight deficiency is that there are some less pleasing aspects of nuclear power plants that we can't ignore. For example, unsightly cooling towers, and difficulties in handling the small amounts of unfortunately highly radioactive waste produced both take a little gloss off what is otherwise a triumph of good engineering design and execution. Therefore, I give the world's first commercial nuclear power plant a score of 3.5 out of 4 in the category of elegance.

Genesis: Whereas the EBR1 scored only 2.5 for novelty, focused on the nature of the solution system, for genesis, which looks more closely at the problem that is being solved, the nuclear reactor scores better. As I stated earlier, this is because the EBR1, to a certain degree, creates as many problems as it solves, so that it can't be judged more favourably as a solution. However, for its effect on the problem space, and our understanding of what needed to be solved, it has some strong qualities. Basically, this means that a solution can be somewhat lacking in novelty itself, but may nevertheless open our eyes to what it is that we are actually trying to solve. The EBR1 seems to be such as example. It is undoubtedly a basis for further, novel approaches to power generation, and represents a new way of looking at the problem of power generation. It also hammers home some of the associated problems of power generation—especially how we generate cheap and plentiful electricity, without destroying the planet in the process. That's now evident with coal (although it may not have been in 1951), but events like the explosion and melt-down at Chernobyl in 1986 leave us in no doubt that power generation is about more than just making the electricity. There are related problems to be solved, like how to deal with the unwanted by-products. The only major weakness in this category is that the EBR1 probably tells us little about solving those unrelated problems. However, as a major disruptor of the power industry, for better or for worse, the EBR1 must score quite well for genesis, and I give it a 3 out of 4 in this category.

Total: The world's first commercial nuclear power plant, built around the Enhanced Breeder Reactor 1, achieves a score of 13 out of 16 for creativity, placing it just in the very high range. It is highly functional in the sense that it is both effective and elegant—it does what it is supposed to do, and it is well executed—however, we have noted some weaknesses in novelty that I suggested result from the fact that it both solves some problems, but introduces others. Despite that slight deficiency, it was quite radical in concept, and helps to break an old paradigm of power generation, that, in current times, may yet lead to the eventual replacement of all dirty and non-renewable forms of power generation, with cleaner and greener renewables. As unpopular as it sometimes is, nuclear power may be the power generation that we had to have, in order to move from non-renewable to renewable energy?

Sputnik I (1957)

The truth is that the US went apeshit – John Naughton – Irish Academic, Journalist and Author (1946-)

Lonely metal souls in the unimpeded darkness of space… Haruki Murakami – Japanese Writer (1949-)

The successful launch of Sputnik 1, on 4 October 1957—the world's first artificial satellite—has a special place in the history of creativity. What became known as the *Sputnik Shock* (see, for example Dickson, 2001) was not only a significant feat of human problem-solving, but it actually gave rise to the modern era of creativity research. In the aftermath of Sputnik's very public achievement, Western Governments, and especially, the US, cast around for an explanation. The reason for the Soviet success was very quickly attributed to the creative prowess of Soviet engineers, and this kick-started our modern interest in the psychology of creativity.

This energy-handling system (we could make a strong case here that it is, or is also, an information-handling system) was 58 cm in diameter, with 4 long, external antennae, and had one simple function: to demonstrate to the world that the Soviet Union had won the first leg of the Space Race. What needs did Sputnik satisfy? At a national level, there is no doubt that Sputnik satisfied an esteem need, and the prestige and sense of accomplishment of Sputnik must have served a useful political and social purpose within the Soviet Union. The Soviets themselves, no doubt would have argued that it also served a vital basic needs, seeing Sputnik as the first step on a path of ensuring their nation's security. More broadly, Sputnik also addressed a global need for self-actualisation, demonstrating, like many inventions before and since, what humankind was capable of doing (Fig. 2).

Fig. 2 Sputnik I, exploded view (*Credit* Wikimedia)

What Was Invented?

Sputnik I (more formally, *Prosteyshiy Sputnik-1*, Russian for *Elementary Satellite 1*) was a very unremarkable device, certainly externally, but even internally. This polished metal sphere, made from an aluminium–magnesium–titanium alloy, and about the size of a beach ball, or a large balloon, weighed approximately 84 kg, and had one single, official function—to broadcast a radio signal as it completed a series of low-earth orbits, for as long as its batteries would last. Sputnik I orbited the Earth at approximately 29,000 km/h, varying in altitude between 223 and 950 km (due to the elliptical nature of the orbit), and completed its orbits in about 96 min.

It succeeded in broadcasting its radio signals for 3 weeks, and continued orbiting for a further 9 weeks, before re-entering the Earth's atmosphere on 4 January 1958. During this 3-month journey, Sputnik I completed 1440 orbits, and travelled about 70 million kilometres.

Although Sputnik's nominal function was to broadcast a radio signal from low-earth orbit, the real purpose, of course, was strategic and geopolitical. The Cold War that had developed at the conclusion of WW2—Germany physically split into two, with half under the control of the Western Allies (the United States, Britain and France), and half under the control of the Soviet Union—and had escalated through the Korean conflict (1950–53). The United States had developed, tested and used atomic weapons in 1945, and the Soviets soon followed with their first successful test of a nuclear bomb in 1949. The German success with ballistic missiles in WW2 (the V2 rocket) had demonstrated the feasibility of combining a missile with a nuclear warhead, and both major powers were racing to be the first to develop such weapons. The urgency undoubtedly resulted from the realisation that such weapons offered the possibility of attacking an enemy without, so to speak, leaving home. They could be launched from a country's own territory, travelled via space, so were virtually untouchable, and threatened to disrupt the balance of power in the Cold War. Sputnik's real role, therefore, was to demonstrate to the world that the Soviet Union had a nuclear ballistic missile capability, and had it before anyone else!

Not surprisingly, Sputnik I had the desired effect, and it is no understatement to say that Western countries, and the US in particular, freaked out as a result. As expected, the Soviet Union had a ballistic missile capability in place by 1958, and although the US followed with their own in 1959, there was a period of great fear and concern in the West, while the Soviet possessed, for a short time, this unique military capability.

It goes without saying that there have been many highly beneficial, and non-military, applications of space and satellite technology since the days of Sputnik I. Global Positioning System (GPS) satellites are familiar to everyone, and accessible through our modern smartphones. Weather satellites help to predict global climate conditions in ways that are used for a range of applications, from farming through to disaster relief. Communications satellites allow us to exchange information in ways, and at speeds, that were never possible before Sputnik. However, as is often the case, and rather unfortunately, it was a military problem—the question of developing nuclear missiles—that was the real driver behind this first satellite.

Why Was Sputnik Invented?

The more of these assessments we do, the more we seem to find that there is rarely a single, underpinning problem, at least for a significant proportion of our inventions. Was the problem how to place an artificial satellite into orbit, or, was it how to embarrass the United States? Or indeed, was it really a question of how to create an invulnerable weapon of mass destruction? All of these are candidates for the driving impetus behind Sputnik, and all can be regarded as addressing the statement "if Sputnik is the answer, what was the question?" Without restricting ourselves to any one of these interrelated problems, let's now consider the creativity of Sputnik.

How Creative Was Sputnik?

Relevance and Effectiveness: It is tempting to think of Sputnik's effectiveness only in terms of the feat of engineering. It is also a rather complex solution to consider, because the satellite itself was only one part of a larger, geographically dispersed system comprising a rocket, ground stations and so on. It is also true that Sputnik was highly relevant and effective for another key purpose—to demonstrate to the world that the Soviet Union was the first country to place a satellite into orbit. Accordingly, it is impossible to give this innovation anything less than the full 4 out of 4.

Novelty: Sputnik fares slightly less well when we analyse its novelty. The key questions here relate not just to originality and surprise, but also to the real purpose of Sputnik. What I mean again relates to the question of Sputnik in purely technical terms, and Sputnik as a propaganda statement. The fact is, the concept of satellites—small, human-made objects placed into earth orbit—was not a new idea, and indeed, both the US and the Soviets were hard at work in this endeavour. Similarly, rockets capable of transporting objects into space, and potentially into orbit around the earth, were not new. The Germans had achieved this in 1944 with their V-rockets. As with many of our innovations, the Sputnik system represented the culmination of a process of incremental development of a particular technology. There is no doubt, however, that the launch of Sputnik, and the highly public nature of its successful placement into orbit, surprised the heck out of the United States! For this reason, I score Sputnik's novelty at 2.5—this is more for its political impact, than its technical qualities.

Elegance: When we turn our attention to elegance, I think Sputnik fares very well. It also does so both for its technical qualities, and for its political purpose. The Soviet space program had many difficulties, often reflecting a lack of access to advanced technologies and other resources, however, it is also frequently used as an example of what can be done under conditions of constraint. The Soviets were able to do more or less everything the US could do, but on a far smaller budget. The joke is that, faced with the need for a device that could write in zero gravity, the US spent millions of dollars developing a special space pen. The Soviets simply used pencils![1] Nevertheless, Sputnik I and its associated systems were skilfully executed, well worked out, nicely proportioned and fitted together logically and functionally. Not only that, but as people all over the world stood on street corners and tuned

---

[1] In fact, this story is unfair to the US. The fire that destroyed the Apollo I capsule and killed its crew during a test highlighted the dangers of using anything in the capsule that might cause possible electrical short circuits—i.e. particles of highly conductive graphite. This is why the US developed *pens* that could write in zero G.

transistor radios into the correct frequency, everyone could hear the beeps emitted by Sputnik that unequivocally broadcast the Soviet's success. I give Sputnik I 3.5 out of 4 for elegance.

Genesis: To what extent did Sputnik I change the paradigm of the space race? To what degree was it a radical departure from previous efforts in the same area? I have already argued that Sputnik was, at its time, the pinnacle of incrementation. It tackled the problem of rockets and satellites the best that it had been tackled up to that point in time, but there is no doubt that it did not even do so the best this could possibly be done—the subsequent history of the space race would demonstrate this. This suggests that there was little that was radical and disruptive about Sputnik. However, in any complex system, there is the overall effect that is achieved, often brought about by lower level elements of disruption in the components. Sputnik also brought attention to previously unnoticed problems and needs—for example, it demonstrated the possibility that a foreign power could reach, unimpeded, into another country, both literally (e.g. with nuclear warheads) and figuratively (through radio broadcasts). Technologically, Sputnik itself was not particularly disruptive, but in terms of what could be done, it opened up a whole new understanding of the potentials of space for military, political and commercial purposes. I, therefore, score it 2.5—perhaps a little low—reflecting the non-technical aspects of the system, more than the purely technical.

Total: 12.5/16—This score places Sputnik at the boundary between high and very high on our scale of creativity. I think this is a fair reflection of the fact that technically, it was primarily the speed with which the Soviets got a satellite into space, rather than the outright technical prowess. It was inevitable that one of the two Superpowers would have achieved this feat around this time. Both were capable, and several other countries were not far behind. What bumps up Sputnik I's score against this criterion is the shake-up that it gave the world. Nobody, except perhaps the Soviets themselves, realised that they were as capable as they were, and Sputnik I well and truly shocked the Western world.

Antisense (Gene) Therapy (1978)

A solid foundation in genetics is increasingly important for everyone—Anne Wojcicki, American Entrepreneur (1973-)

It may seem somewhat counter-intuitive to refer to a medical treatment as an information-handling system, however, that seems to be the best description of an invention based on DNA and genetics. If we ignore the fact that we are talking about biological systems, the idea of removing and replacing defective units of genetic information—editing them in a sense—sounds comfortably like a form of information processing. Antisense therapy may also occupy an interesting place as a solution to some of Maslow's needs. It seems clear that this technique addresses basic, physiological needs. However, does it also not address some of the highest needs—if we succeed in curing disease in an individual, is that not a necessary precondition for that individual achieving their full potential? Let's now look more closely at antisense therapy.

What Was Invented?

Most people nowadays have at least heard of DNA—deoxyribonucleic acid—and have some notion of it as one of the building blocks of life. DNA molecules hold all of the genetic information required by organisms to grow and reproduce. The double-helix structure of DNA (see Fig. 3), discovered in the early 1950s thanks to the X-ray imaging work of Rosalind Franklin, and the subsequent research of James Watson and Francis Crick, has been the foundation of many important advances in medicine.

Once scientists understood the nature and structure of DNA and associated molecules, and their role in biology and genetics, medical applications soon followed. *Gene Therapy*, in its broadest sense, encompasses a variety of different ways that scientists and doctors seek to treat disease through genetic manipulation. First conceived in the early 1970s, the most common form of gene therapy, which was successfully trialled in the early 1990s, involves inserting new *therapeutic* genes into an individual in order to replace defective, *mutated* genes—rather like a transplant, just on a much smaller scale. A subsequent advance, however, came in the form of blocking, or deactivating, genes that were already present in the body, in order to disable them, and therefore prevent some undesirable biological outcome such as a particular disease.

Paul Zamecnik (1912–2009), as a Professor of Medicine at Harvard Medical School, invented *antisense* therapeutics in 1978. This is an example of the latter type of gene therapy, involving switching off a gene to prevent it from causing a disease. It has to be said that the whole field of gene therapy in medicine is highly complex, and, like all medical research often proceeds very slowly while extensive human trials are conducted. Although invented in 1978, there are only a small number of proven antisense treatments, with a modest number of others undergoing clinical trials. Examples of diseases for which there are available antisense therapies include *cytomegalovirus retinitis* (an inflammation of the retina that can lead to blindness) and *spinal muscular dystrophy* (a neuromuscular disorder that causes muscle wasting and frequently leads to premature death). There are many more antisense therapies undergoing trials, including for diseases such as Ebola and AIDS. This technique holds great promise and, despite the complexity and challenges involved, may yet yield effective treatments to many more common and debilitating diseases.

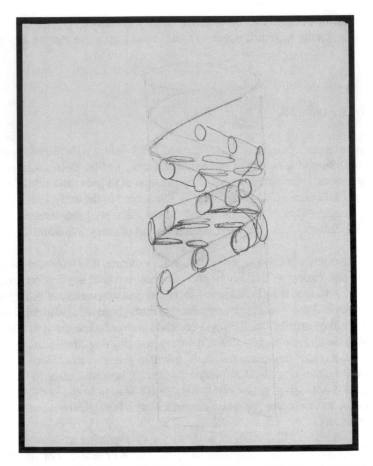

Fig. 3 Pencil sketch of the DNA double helix (*Credit* Wellcome Collection, Francis Crick)

Why Was Antisense Therapy Invented?

As frequently seems to be the case, our inventions address not only a very specific problem/need, and also a more general, but related need. Let's give these some scientific names and say that the *proximate* problem—that is, the very specific problem at hand—in the case of antisense therapy is *how to replace defective genes*. The *distal* problem—the bigger picture—is *how to cure disease*. These are related, and the proximate problem is, and should be, just a more refined and specific expression of the distal problem.

How Creative Was Antisense Therapy?

Relevance and Effectiveness: I feel the same guilt that I commented on in my assessments of the smallpox vaccine (which I scored 12/16) and antibiotics (which did somewhat better on 13.5/16) as I give you my assessment of antisense therapy. Probably like many people, our instinct is to rate medical breakthroughs very highly. Most people alive today have benefited in some way from the medical inventions and discoveries of bygone eras, and there is no doubt that we will marvel at other medical breakthroughs that are yet to come. However, as always, I try to take a cold, hard and objective view in making these assessments of creativity, and antisense therapy comes up a little short in this category. The technique and the practical examples of it are technically and biologically correct. They are a culmination of the application of a body of knowledge and expertise. The main difficulty in this category is that they are, in general, not terribly effective. If you are suffering from spinal muscular dystrophy, and face the certainty of disablement and premature death, of course, a treatment that is even 50% successful is worth every penny. What I mean by poor effectiveness is not the effectiveness of a specific antisense treatment, which may work impressively well. Rather, I mean the effectiveness of the technique in general. For a technique that was invented 40 years ago, and looking at it as a solution to a problem, I feel that we have to rate its performance as somewhat deficient. I'll be the first to admit that I was wrong when I'm benefitting from some future, life-saving, antisense treatment, but against our creativity criteria, I stand by the assessment. Antisense therapy scores 2.5 out of 4 for relevance and effectiveness.

Novelty: In contrast to relevance and effectiveness, antisense therapy scores highly against the indicators of novelty. Not only does the technique help to define the problem—how to stop bad genes doing bad things to our bodies—but it extends the whole understanding of DNA, genes and associated elements in a new direction. This treatment method showed medical scientists a new way to apply their knowledge of genetics, resulting in new treatments for diseases. Antisense treatment also highlights the shortcomings of other, conventional approaches to the treatment of disease, and we can look at this approach as a major improvement over older techniques whose effectiveness had reached a limit. Antisense therapy scores 3.5 out of 4 for novelty.

Elegance: So often we have observed the relationship between elegance and effectiveness—as one increases, so does the other. I believe that there is a direction to this—elegance is the *cause*, effectiveness is the *result*—and as soon as we see a drop in effectiveness, we can usually predict that elegance (or lack thereof) will be a factor. That is also the case here. At least some of the weakness we identified in the relevance and effectiveness of antisense therapy is a result of the complexity of this technique. Individual instances of treatments may be well executed, and even the general concept of switching off or replacing genes is quite simple and self-evident, but the practicalities of the technique, and the specifics of developing treatments demonstrate an underlying complexity. Furthermore, it may be that this inherent complexity simply makes it hard to develop a successful antisense treatment. That difficulty and complexity then raises the prospect of a given treatment failing to work, and ultimately, contributes to a lowering of relevance and effectiveness. Put all of this in a different way, if the development of antisense therapy was easy, there would be far more examples of successful antisense treatments! None of this makes this innovation bad, it just highlights the fact that there are challenges inherent in solving the underlying problem. Against the criterion of elegance, I score this invention 2.5 out of 4.

Genesis: Antisense therapy, like some of our other inventions, oscillates back and forth in our creativity criteria. It's weaker in one element, but stronger in another, then weaker again, and finally, we finish on a strong note. As we have seen now many times, genesis asks us to look at the invention in question in terms of its impact on a field, and how it causes us to understand the problem. Antisense therapy is part of a general shift in medicine that breaks away from an established paradigm. In fact, we have seen this sort of shift more than once throughout history. Once upon a time, people believed that diseases resulted from an imbalance of the four humours. Blood, black bile, yellow bile and phlegm, and their relative proportions in our bodies, were thought to control health and emotions. Consistent with this model, the treatment of diseases was a matter of eliminating an imbalance, for example through the practice of blood-letting. As part of the general field of genetics, antisense therapy represents another (and far more effective) paradigm shift. In many instances, existing treatments of the disease have (or had) reached their limit in terms of what could be done to address the underlying cause. Genetic therapies opened up a whole new class of treatments of a fundamentally different type, and antisense therapy is a specific instance of this. Not only that, but this innovation lays out a path for further new work, and continues to cause medical researchers to look at diseases, and their treatment, in new ways. Antisense therapy establishes new benchmarks against which we judge treatments, and for these reasons, as well as its positively disruptive effect in the field of medicine, I score it at a strong 3.5 out of 4 for genesis.

Total: Antisense therapy scores 12 out of 16 for creativity. This places it in the high range for creativity. There are two barriers keeping it outside of our highest classification, and these relate not to the novelty or the ground-breaking nature of the invention, but to matters of practical execution, effectiveness and complexity. Although we may feel somewhat dissatisfied with a score of only 12, the particular makeup of this score gives me great hope, because the pathway to the maximum score of 16 is clear and open. As the technique improves—as medical scientists find ever better and simpler ways to execute antisense therapy, I think we can expect to see more of these treatments becoming available. The technique, in many respects, is still in its infancy, even though it has existed for around four decades. Perhaps that is also a reflection of its complexity, and may be unlike other areas of technological innovation. However, it seems clear that we haven't heard the last from this important medical innovation.

The Digital Age (1981–Present): The Rise of Complexity

Our final epoch is the one that we currently occupy. Although this spans the period from the 1980s to the present day, historians may eventually differentiate between the early Digital Age—running from the advent of the Personal Computer (PC) in 1981, to about 2015—and a post-Digital Age. In technological terms, the latter (or current) part of the Digital Age—what I am suggesting may eventually be recognised separately—is characterised by a new Industrial Revolution, and one that has been labelled *Industry 4.0*. The First Industrial Revolution is the one that we covered under the Age of Enlightenment. The Second Industrial Revolution was characterised by mass production, assembly lines and electricity, and corresponds approximately to the late Romantic Period, and early Modern Age. The Third Industrial Revolution is defined by the rise of computers and automation—intersecting with the late Modern Age and early Digital Age—but the Fourth Industrial Revolution (Industry 4.0) represents the synthesis of data analytics, autonomous systems and artificial intelligence in what is termed cyber-physical systems.

The first invention we will tackle is another medical innovation. Artificial skin was developed at the beginning of this period, in 1981, and I have included it here because it is a fascinating example of a modern, but fairly low-tech (in an Industry 4.0 sense) solution. There are no AI, and no electronics in this invention, although modern IT no doubt was, and is, vital to the development and use of this innovation. What artificial skin shows us, apart from another in a long line of medical innovations across the centuries, is the challenge we face when trying to solve many different problems with a single solution.

The remaining two inventions that we consider here are both key contributors to, and representative of, the rapid transition to Industry 4.0. The first is the ubiquitous *World Wide Web* (WWW). Like the Industrial Revolution, or sewerage systems, it is more *system* than product, and the web is an example of a market-driven solution (a communications tool for researchers) that also keeps delivering new solutions that are looking for a problem. We appear to be nowhere near the end of the potential uses of the WWW. The last of our Industry 4.0 innovations is the *smartphone*. Rather like the WWW, the smartphone is proving to be both a pinnacle of technological achieve-

© Springer Nature Singapore Pte Ltd. 2019
D. H. Cropley, *Homo Problematis Solvendis—Problem-solving Man*,
https://doi.org/10.1007/978-981-13-3101-5_12

ment—*look what we can do*—but also a driver of further technological change. There seems no end to the applications of these technologies.

Artificial Skin (1981)

Huge sums are invested globally in medical research – and with good reason, Geoff Mulgan CBE, British Chief Executive of NESTA, (1961-)

The first innovation in our final era is a material-handling system. That might seem like a curious designation for artificial skin, but apart from the fact that it is clearly not an information-handling system, and not an energy-handling system, this innovation is designed to replace the function of a human organ, and allow that organ to regenerate. What need is addressed, in the sense of Maslow's hierarchy? Clearly, a basic, physiological need is tackled, providing a patient suffering from severe burns with a protective barrier that improves their chance of returning to good health. Artificial skin, however, also addresses a psychological need. As anyone who has suffered a disfiguring injury will confirm, there are psychological scars that must also heal, and artificial skin plays an important role for this higher order need as well.

What Was Invented?

Throughout the course of this book, I have emphasised the connection between change, problems and creative solutions. Change, you will recall, is the key. It can either be the driver of new demands—sometimes referred to as *market-pull*—or it can be the stimulus for new inventions—often referred to as *technology-push*. Market-pull defines new problems, while technology-push defines new solutions, and the process of linking these together *is* innovation. In this book, I have tried to show this link through each invention, not only by telling you about the invention (i.e. the solution), but also by highlighting the problem that it solves (succinctly expressed in the form *How to Verb Noun*).

Sometimes, the underpinning problem is relatively simple (though not trivial)—e.g. *how to keep time*. In some cases, however, the underpinning problem is more complex, and artificial skin seems to be a good example of this complexity. The challenge seems to be that the problem is not a single issue: e.g. how to make an artificial skin. Rather, it is really a *suite of related problems*. These related problems are all of the things that real skin does—all of its functions—which artificial skin must replicate if it is to be a true solution to the problem. Putting this another way, artificial skin actually solves multiple problems simultaneously, and the more problems it has to solve, the harder it is to come up with one single, integrated solution. However, in 1981, Massachusetts General Hospital surgeon John Burke, in collaboration with MIT chemist Ioannis Yannas, succeeded in producing an artificial substance that,

Fig. 1 Artificial skin (*Credit*
https://www.dreamstime.
com)

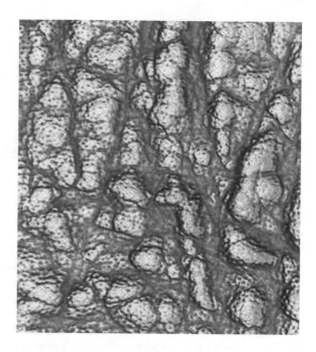

like real skin, was a barrier to infection, prevented dehydration and provided a temporary structure around which a patient's own new skin cells could grow. Burke and Yannas's artificial skin was particularly vital for patients, who had suffered burns to more than 50% of their body, for whom traditional skin grafts were not a viable option.

The artificial skin developed by Burke and Yannas used naturally occurring collagen fibres—essentially an elastic protein—taken from cowhide, and combined these with long-chain sugar molecules found in shark cartilage, and created a polymer membrane which they attached to a plastic sheet to create artificial skin (see Fig. 1).

Why Was Artificial Skin Invented?

In the previous section, I noted the particular complexity of artificial skin—it is not really a single problem, but a set of related problems. For this reason, rather than expressing the problem as the obvious *how to create artificial skin*, it is probably more meaningful and relevant to express the problem in a different way. There's another lesson of good problem statements that we should remind ourselves of as well. Good problem statements should be solution free. In other words, they should not constrain the form of the solution. How to create artificial skin is rather restrictive, because the only possible solution is artificial skin. If we state the problem in a more solution-free form, we leave the door open for greater creativity. In other words, the artificial skin should be just one possible solution to the stated problem—which I think here is better expressed as how to replicate real skin?

How Creative Was Artificial Skin?

Relevance and Effectiveness: When I first started writing these case studies, I assumed that every one of our inventions would, almost by definition, score 4 out of 4 in the category of relevance and effectiveness. However, we have seen a number of cases where the performance of the innovation in question has been slightly deficient. This makes sense, once we have the benefit of reviewing these inventions as a set. Some have been only partial successes, and have fallen some way short of solving the problem they were designed to tackle. However, this has rarely been a major impediment, and the inventions that have scored less than the maximum, have usually served simply to stimulate further innovation. In many ways, this highlights the fact that innovation is a never-ending process. No invention ever really completely solves a problem because our old friend *change* is not static. Even an apparently perfectly functional solution to a problem usually becomes obsolete, not least because market-pull—the customer, or user—demands something better. As soon as the automobile was invented, people wanted better automobiles! The inventions that we have considered with a score of, say, 3.5 out of 4 in this category, including this current invention, are just another stepping stone on the path of innovation. Like a mirage, we never quite reach the goal, but that is one of the factors that keeps innovation, and invention, a thriving business.

Novelty: Another pattern that has emerged in our analyses is a trend towards small deficiencies in *diagnosis*—the ability of an artefact to draw attention to shortcomings in other inventions—and *prescription*—the characteristic of an invention to show how other artefacts could be improved. It seems to make sense that any new, effective artefact should simultaneously highlight what is wrong with other competing solutions, and give hints as to how they might be made better. In the case of artificial skin, we see this trend again in operation. The deficiency in scores here is due to the fact that our example of artificial skin itself embodies some weaknesses. For example, it does not address the problem of allowing heat to be lost from the body through sweat because it does not include sweat glands, and it does not solve the problem of allowing hair to grow. Therefore, we can't score this innovation at a maximum, because it is, itself, somewhat deficient and open to improvement. Similarly, I feel it has some weakness in what it offers in terms of new and different ways of using the artefact. Some inventions spark off all sorts of other ideas, but in the case of artificial skin, it is harder to imagine other uses for the material. Perhaps this weakness lies in the very fact that it solves several problems in one. The more the invention sits at the intersection of several sub-problems, the more limited its alternative uses may be. I score it 3 out of 4 for novelty.

Elegance: There is a particular danger in passing judgement on the creativity of some inventions, because some of these innovations evoke stronger feelings, and solve problems that are, literally, matters of life and death. I have already rated one of history's great life-savers—the smallpox vaccination—a mere 12 on our scale, while Antibiotics fared a bit better on 13.5. A fair criticism of my scoring might be "if you think you can do better…" However, my defence is that progress is stimulated by change, and constructive criticism is a form of market-pull! Hence, I'm going to continue to criticize artificial skin, even though I am sure it has been of immense benefit to the lives of many people. In the category of elegance, I scored this invention 3 out of 4. My reasoning is that there remains room, with this invention, for improvements in execution. A more elegant—that is, a better executed solution—might be one that is sprayed on, instantly and permanently replacing the natural skin that has been lost. Again, that might seem a harsh criticism, but that is how innovations continue to evolve.

Genesis: Although I scored artificial skin at only 3 out of 4 for Novelty, it does somewhat better in this category. At first glance, that seems a little strange. However, this may highlight the interplay between these two related categories. There is a certain internal and solution-focused nature to novelty—how novel is this invention, and how does it compare to other inventions (addressing the same problem). Genesis, on the other hand, is more focused on a bigger picture. What is the impact of this invention, and also, how does it affect the way we understand the problem. In that sense, artificial skin can be less than completely novel when judged against other similar inventions, but quite ground-breaking in terms of how it addresses the underlying problem. Artificial skin has a strong degree of foundationality—it breaks the existing paradigm of borrowing skin from elsewhere on the patient, and changes this to a paradigm of replacing normal skin with some substitute. Thus, the problem is no longer how to graft healthy skin onto a damaged area, but how to attach artificial skin as a temporary substitute. Another way to see the impact of genesis, in this case, is to recognise that the skills and knowledge required by doctors using artificial skin are considerably different from those used in grafting skin. We can imagine that surgeons had to undergo some specialised re-training, and this reflects the disruptive impact of an invention with a high degree of genesis. I, therefore, scored this invention 3.5 out of 4.

Total: The invention of Burke and Yannas has a total creativity score of 13 out of 16. This places it in the very high range of our continuum. The principal weaknesses that detract from its score are a slight deficiency in effectiveness—there are some core functions of skin that it did not address—and a certain lack of novelty and elegance. However, I have already suggested that these deficiencies often serve as the stimulus for further improvements. Like many of our inventions, if we look ahead in time, we see that many improvements have been made, such that a score of 12 or 13 for creativity may indicate just enough creativity to be a valuable solution, but just enough room for improvement that led, or will lead, to further developments of real value.

The World Wide Web (1989)

...a relatively simple invention, with profound social and economic consequences... - Katie Hafner, American Journalist (1957-)

The World Wide Web may be the ultimate information-handling system. Driven by the possibilities offered by hardware (computers and computer networks), and stimulated by a desire to exchange scientific information amongst researchers, the WWW had its origins as a means of satisfying some of our highest, self-actualisation, needs. Humankind has always shared information, even before the invention of writing, of paper, and of the printing press, however, we have been driven to find ever better ways to share our ideas, knowledge and innovations. Electricity, computer technology and other inventions have continued this progress, but the Web probably sits at the pinnacle of all of these. Of course, an invention's impact, or potential, is not the same as its creativity, and you may be slightly surprised by the score that I give it. That takes nothing away from the contribution that the Web is likely to continue to make to the development of humankind, and our ability to find all sorts of other solutions to our needs and problems.

What Was Invented?

The World Wide Web was invented by English engineer and computer scientist, Tim Berners-Lee, in 1989 while working for the European Organisation for Nuclear Research (known by its French acronym CERN). The Web was created to be an information space using the physical resources of the Internet[1] to exchange documents, images and other digitised media among computers.

Many of the necessary components of the Web existed prior to Berners-Lee's invention, but it was his particular integration of two elements that gave rise to the now familiar *Web*. Hypertext is a system of representing text (i.e. writing) on computers and other electronic devices, with the additional feature that multiple levels of detail can be progressively accessed (e.g. with a mouse), and where other text can be accessed, all through the mechanism of *hyperlinks*. The hypertext/hyperlink concept is thought to have existed in some form since as early as the 1940s. The Internet, in its strictest sense, is now the global system of interconnected computers and computer networks and traces its origins to systems first created in the United States in the 1960s (e.g. ARPANET). Berners-Lee's innovation was to join these two systems together, in the process creating three key enabling technologies: (a) the Hypertext Transfer

---

[1]As ubiquitous and familiar as the things like Google and the World Wide Web now are, it's important to understand the nature and functions of the different parts that combine together to give us what is typically referred to nowadays as *The Internet*. In very simple terms, The Internet is the physical system of globally interconnected computers, on which the World Wide Web resides. The Web itself is a system of rules, language and standards, that allows documents, images and other media to be shared and viewed on browsers connected to the Internet.

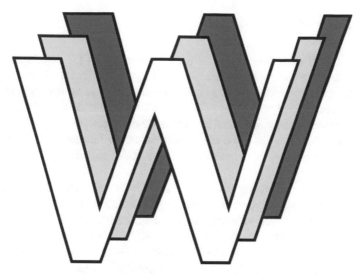

Fig. 2 Black and white WWW logo, 1990 (*Credit* Wikimedia, Robert Cailliau, Hell Pé)

Protocol (familiar to us from the HTTP in web addresses); (b) the Hypertext Markup Language (HTML) that was necessary for writing and publishing web pages, before the days of smarter development tools, and; (c) the system of unique identifiers for materials on the Web, now familiar to us as the Uniform Resource Locator (URL)—what we commonly call a website's "address". These three elements created a uniform, scalable and consistent means for sharing content over the Internet. Prior to the Web, computers could send digitised information to each other over networks, but this process was far less accessible, and far more specialised than it is now. The Web provided everyone with a single language, and a single, unified *map* for the vastness of the Internet (Fig. 2).

Why Was the World Wide Web Invented?

The difficulty with the Web is that we are in the midst of experiencing its impact. This makes it even harder to assess it with a clear perspective, because it's apparent that the full impact has not yet been realised. What we do know is that Berners-Lee's aim was to create a global, hyperlinked, information system—a way of sharing documents (and other media) that was easily accessible. What he probably didn't anticipate was the extent to which the system he created would become so ubiquitous and influential. Many of the things that the Web now does have emerged since the basic concept was implemented in the early 90s, and many more uses continue to emerge, but the clear, driving problem that gave birth to the Web was the simple question of *how to share information?*

How Creative Was the World Wide Web?

Relevance and Effectiveness: I will give the Web a score of 4 (the maximum) for this criterion. The immense growth of the Web over a period of barely 30 years, and the proliferation of applications that have been developed, leave no option but to give it the highest score for relevance and effectiveness. Of course, it could be argued that there are undesired consequences of the Web as well—fraud, pornography, bullying, to name just a few—but the good of the Web seems to far outweigh the bad. As a means of sharing information across the world, it is unrivalled in its reach, and in its effectiveness.

Novelty: As has frequently been the case, this is somewhat difficult to judge because there is a perspective based on hindsight, as well as an attempt to assess how novel any innovation appeared when it was first launched. The incremental nature of the Web suggests that its novelty was not at an absolute maximum—it's likely that some sort of Web-like innovation would have appeared at around the same time, and it is a logical extension of the resources and technologies that already existed at the time it was launched. A standardised way of communicating over the Internet that existed in the late 80s/early 90s cannot be seen as a paradigm shift, but the Web was new in a narrower sense. For this reason, I give the World Wide Web a score of 3 out of 4. This score reflects the largely incremental nature of the Web's novelty: it was a new solution to an existing problem, so it achieves a high, but not perfect, score for novelty.

Elegance: As a complete, fully worked out, solution, the Web fares remarkably well. If we look at the actual development process and time it took to implement, the Web was established quickly, and with surprisingly little difficulty, suggesting that Berners-Lee's solution was well designed, and well executed, from the outset. The fact that the basic system has also existed, with little real change, for nearly 30 years, reinforces the sense that the Web was a highly *elegant* innovation. I give it the maximum score of 4 in this category.

Genesis: Was the Web a new solution to a new problem? No! I have already argued that it was largely an incremental response—a better way of solving an existing problem—and therefore, cannot be seen as a radical, paradigm-breaking innovation. I must admit that I feel uncomfortable writing that! It seems that the Web should be super-innovative in a radical sense. However, applying the cold, hard lens of objectivity, it is difficult to justify high scores for genesis—it was a new solution to an existing problem, not a paradigm-breaking, problem redefining new solution to a previously unrecognised problem. For this reason, I see no option but to give it a score of 2 here.

Total: The WWW has a total score of 13 out of 16, which places it the very high range. The principle weaknesses that hold the Web back from one of the highest scores in our set of inventions relate, as has frequently been the case, to the incremental nature of the innovation. Indeed, if there is a theme that has emerged from this analysis, it may be the fact that very few inventions—in the sense of deliberate solutions to identified problems—are truly radical. This does not mean that radical innovations—true paradigm breakers—do not exist. Rather, I think it reflects the fact that there are two pathways to innovation. The first, which has been our principal focus, begins with a need, and ends with a solution. The other pathway begins with a solution, to which a need must then be attached. I argue that the former is the basic paradigm of invention, while the latter may be more characteristic of discovery. Both, of course, connect solutions to problems, but they do so in different ways. Discoveries typically emerge with little prior warning, and therefore lead to problems that have not yet been anticipated. Because inventions respond to problems, there is always some degree of anticipation of the solution. An invention can never be truly surprising, because the act of defining the problem anticipates some sort of solution. In the light of this discussion, it is tempting to suggest that, on our scale of creativity, it is impossible for an invention to achieve a score of 15 or 16, and that a score of 13 and 14 represents the pinnacle that can be achieved. Of course, we have already seen inventions that exceed 14 on our scale! Regardless, the WWW is highly creative, and occupies a high position in our ranking of inventions.

The Smartphone (1992)

I can't live without my iPhone – Sara Paxton, Mexican–American Actress (1988-)

We have reached our last invention! This information-handling system has had a remarkable impact on society in a lifespan of barely 25 years. For many of us who are old enough, it is hard to remember how we managed without one of these devices in our pocket. For people who have grown up with the device, the idea that there was a time when you were not in constant contact with the world may seem hard to imagine. It is interesting, nevertheless, to try and reconstruct the need that was satisfied by the first smartphones (and indeed, the first mobile phones). Was it a basic need—for safety and security—that drove the invention of the smartphone? Was it an unfulfilled need to stay in contact with friends and loved ones? Was it a need to demonstrate to the world your status and importance—look at me, I have a smartphone! Or, was it because the smartphone was the means for finally achieving our full potential as humans? If I sound a bit cynical, it's because, at the time, my feeling is that none of these needs (except perhaps the need to show off) was crying out for the smartphone as a solution. And yet, now that we have them, there are undoubtedly many impressive benefits to be derived. What this suggests to me is that the smartphone was probably a good example of technology-push—a solution waiting for a need. Happily, it seems that there's no end to the problems that we find we can solve with this clever device.

What Was Invented?

There is an important distinction to make here; one which may be obvious to people like me (Gen X), but less obvious to Millennials and other so-called Digital Natives, who may have grown up knowing only the smartphone. I am very deliberately referring to the smartphone as distinct from mobile or cell phones. By the latter, I mean the more mundane, and less capable, telephones of the 1980s and 1990s—basically a portable, handheld telephone—and by the former, I mean the integrated devices that combine the functions of telephone and computer, running an operating system and a variety of applications, and which fit into your pocket (Fig. 3).

The first such device began on the drawing board of IBM electronics engineer Frank Canova (1956–). The early 1990s was an era when a number of innovations in consumer electronics began to reach a level of maturity and price that put them in the hands of a growing number of people. Sony had introduced the Walkman portable

Fig. 3 The IBM Simon
personal communicator and
charging base (*Credit*
Wikimedia, Bcos47)

cassette player in 1979, while IBM had launched the modern PC in 1981. Throughout the 1980s, we saw the introduction of other popular devices—the modem in 1981 (allowing computers to talk to each other across telephone lines); the camcorder in 1983 (integrating mobile video technology into a single unit); the laptop computer in 1983 (although you will be horrified to hear that it weighed 11 kg); the Apple Macintosh in 1984 (setting the scene for a vast range of electronic innovations); the touchpad interface device in 1988 (as a practical replacement for the mouse), and, of course, the World Wide Web in 1989.

By the early 1990s, therefore, companies like IBM were under pressure to maintain their leadership and success in this field by making their products smaller, lighter and faster. It was just such an initiative that led Canova to see the potential of the new chip and wireless technology coupled with novel, solid-state memory technology. His concept of an integrated device that blended mobile telephony with functions available at that time on a PDA (Personal Digital Assistant) was dubbed the *Angler*, and was demonstrated at the COMDEX computer and technology trade show in Las Vegas, in late 1992. The Angler, later marketed as the *Simon Personal Communicator*, was sold to the public from August 1994, and was capable of making and receiving calls, sending and receiving faxes and emails, and also included many now familiar features such as a calendar, notepad and calculator. It sold, initially, for US$899 with a 2-year service contract, and 50,000 units were sold in only 6 months, before the device was superseded by newer models.

By modern standards (only 24 years later!) the IBM Simon is slow, heavy, cumbersome, and had a battery that lasted about one hour. Nevertheless, although the term smartphone was not introduced until after the IBM Simon entered the market, this device has become recognised as the grandfather of them all. It is estimated that there will be 2.5 billion smartphones in use in the world by 2019—a far cry from those first 50,000!

Why Was the Smartphone Invented?

I made the case, in the introduction to this innovation, that the smartphone may be best viewed as an example of technology-push. Right at the start of the book, I discussed this in contrast to market-pull. The former, like the smartphone, is a case of the solution coming first, and then connecting to a need. While in general, this is not uncommon, we have focused most of our attention on the need coming first—market-pull—and that then driving the invention of a solution. This situation, however, makes it a little trickier to state the problem that the smartphone was designed to solve. To a large extent, the motivation was simply "because we can", or to demonstrate what was possible with the technology that was emerging at the time. As I described earlier, electronic circuit technology in the early 90s had developed to a point that made it possible to integrate a variety of functions onto chips, and the smartphone was an opportunity to demonstrate what was possible. These kinds of speculative technology developments don't always pay off, precisely because sometimes it turns out that there is no underlying need. However, when they do pay off, frequently they pay off big! If we have to state a problem for the smartphone, then a good one may be simply *how to push the boundaries of technology.*

How Creative Was the Smartphone?

Relevance and Effectiveness: A challenge in making this assessment is to look at smartphones in a general sense, rather than focusing on a single case, such as the iPhone. As a class of devices, smartphones seem to be highly relevant and effective. This is not to say that there have been no missteps in what is still a very new technology. In general, however, the smartphone is highly successful as a solution to the problem of mobile communications and computer applications. Not only do most smartphones do what we expect—allow us to make calls, send texts, take photographs, share content, navigate and now even translate speech—but they fit a basic constraint of handheld portability. Not surprisingly, I give them a maximum score of 4.

Novelty: Smartphones embody a high degree of novelty, and yet I found myself scoring them slightly below the maximum, with a 3. I think this reflects the fact that, to some extent, there is still a lot of incremental creativity in the smartphone—in other words, a lot of their value derives from the fact that they do what previous mobile phones, and other types of computers did/do, only they do it in a smaller, more portable fashion. The fact is, however, that they still bear a strong resemblance to mobile phones, and they function largely like desktop computers. They are a new solution to an old problem. For this reason, I give smartphones the score of 3. Before you protest, read on!

Elegance: Is the typical smartphone elegant? Steve Jobs supposedly said that you know you have the design right when you want to lick it, and many smartphones come close to this! They are, without a doubt, well-executed electronic devices. They are nicely finished, frequently very well designed, considering the hardware and software that must be integrated into such a small platform, and usually nicely formed. I would like to give them a 4, but I also have to look at what detracts from their elegance. They seem to break easily, as anyone who has dropped one has found. They still have a frustratingly poor battery life, limiting their usefulness in some situations. People like me, whose eyes aren't what they used to be, can find the screens hard to read. If we are being honest, they could be designed in a better, more user-friendly way, meaning that I feel they deserve only a 3 out 4, no matter how lickable they might be!

Genesis: If you are still annoyed at me for giving the smartphone a score of only 3 for Novelty, I am about to redeem myself. One thing that I think this innovation does, which is important as a component of creativity, is that it draws attention to previously unnoticed problems. A simple example, notwithstanding my earlier complaint about battery life, is that these devices highlighted weaknesses in battery technology, and have undoubtedly stimulated great improvements. Another important feature of the smartphone is also the fact that it shows some kind of emergent properties—individually, the functions of phone, SMS device, pocket calculator, camera, GPS receiver, are fairly routine—put them together in one device that people carry around with them everywhere they go, and suddenly, interesting and paradigm-breaking things become apparent. Who would have envisaged the possibility of smartphones themselves becoming the source of data for a comprehensive map of traffic flow, congestion and routing? This is the kind of unanticipated outcome that genesis seeks to describe. Therefore, I leaned heavily towards a higher score for genesis—3.5/4—and I think this is rather unusual. There seems to be a general progression from incremental to radical innovation. In the case of the smartphone, it is basically, individually an incremental improvement to an old problem. But put a bunch of smartphones together, and suddenly you have something entirely new—a system of systems, with a set of new and interesting properties!

Total: The smartphone's score of 13.5 out of 16 puts this invention in the very high range, but only just! That comment reminds me that we have to be careful not to assume that the most creative inventions are the most recent. Aside from the fact that the smartphone is rapidly approaching its 25th anniversary, the fact is, we have reached the end of our journey and it is time to cast a critical eye over our collection of inventions!

The Innovation Scoreboard

Now that we have explored our set of 30 innovations across 10 time periods, it's time for some analysis! Back in the introduction, I stated that I was attempting to take a criterion-based approach to the assessment of creativity. There was no predetermined constraint on how many inventions could score a maximum on the CSDS scale. I attempted to assess the creativity of each invention on its own merits, trying to imagine myself present at the time the innovation was introduced, rather than using the benefit of our extensive hindsight.

In fact, the nature and distribution of the creativity scores are interesting and rather unexpected. It was certainly *not* the case that the modern, high-tech inventions score higher than older, and often simpler devices. One immediate observation is that our capacity for creativity and problem solving is a strong and persistent characteristic of anatomically modern humans, constantly rising to new challenges and embracing the technological possibilities of any given era. Let's now look at the ranking of our catalogue of inventions based on total creativity score (out of 16). Please refer to the table in Appendix B for the complete set of scores across all criteria.

1. The Velocipede (16)
2. The Industrial Revolution (15.5)
3. The Crankshaft (15)
4. The Steel-girder Skyscraper (14.5)
5. The Pendulum Clock (14.5)
6. The Slide Rule (14)
7. Cuneiform Writing (14)
8. The Leyden Jar (13.5)
9. The Linen Condom (13.5)
10. Smartphones (13.5)
11. Antibiotics (13.5)
12. The Steam Pump (13.5)
13. Artificial Skin (13)
14. The World Wide Web (13)

© Springer Nature Singapore Pte Ltd. 2019
D. H. Cropley, *Homo Problematis Solvendis—Problem-solving Man*,
https://doi.org/10.1007/978-981-13-3101-5_13

15. Oars (13)
16. Paper (13)
17. The Spinning Wheel (13)
18. Nuclear Power (13)
19. Sputnik I (12.5)
20. Sewerage Systems (12.5)
21. The Movable-type Printing Press (12.5)
22. The Electric Lightbulb (12.5)
23. The Construction Crane (12.5)
24. The Smallpox Vaccine (12)
25. The Julian Calendar (12)
26. Antisense (Gene) Therapy (12)
27. The Hand Axe (11.5)
28. The Wright Flyer (11.5)
29. The Scythe (11)
30. Standardised Coinage (11).

One important factor to keep in mind is that all of the inventions are *creative* in a general sense. We are looking at a very *unrepresentative* sample of inventions, in the sense that I only chose innovations that were inherently new and effective. Of course, some possess these qualities more than others, but the point is that none are *uncreative*. For this reason, it's important to keep in mind that Standardised Coinage, with a score of 11/16, is not uncreative. Rather, it is simply somewhat less creative than the Velocipede (16/16). For an innovation to be pushing the bounds of actually *un*creative, I would say that it would have to score 8/16 or less on the CSDS scale, and in particular, would be expected to have scores for effectiveness and novelty no higher than 2 out of 4.

We can gain some additional insights into the creativity of these inventions by exploring some of the subcategories of creativity used in the CSDS. Because most of our inventions are good solutions to a problem, they score highly for effectiveness. Rather than examining their ranking for that criterion alone, it is more revealing to look at effectiveness+elegance (with a maximum score of 8). These two criteria together could be said to define *functionally* creative solutions, i.e. those that do the job they are designed to, and that are well executed. Let's look, therefore, at the top *functional solutions* in our catalogue:

1. The Velocipede (8);
2. The WWW (8);
3. Steel-girder Skyscrapers (8);
4. The Slide Rule (8).

These represent quite a spread across time periods, ranging from 1622 (the Slide Rule), right up to the Digital Age. They also represent a cross section of energy- and information-handling systems.

It is also interesting to look at the top three innovations by novelty—*the surprisers*—to see if any pattern or features stand out in this respect. The top surprisers, out of a maximum score of 4, are the following:

1. The Velocipede (4);
2. The Pendulum Clock (4);
3. The Industrial Revolution (4).

Here, we find innovations spanning energy- and information-handling systems, from just two adjacent time periods.

Looking only at the criterion of elegance—the *pleasing solutions*—we find a close correlation to the functional solutions. Indeed, we observed, on a number of occasions that effectiveness and elegance seem to go hand in hand. The pleasing solutions, out of a maximum score of 4, are the following:

1. The Velocipede (4);
2. The WWW (4);
3. Steel-girder Skyscrapers (4);
4. The Slide Rule (4).

Finally, it is also interesting to look at the top inventions by genesis—what we can call the *disruptors* or *paradigm breakers*. In this category, out of a maximum score of 4, we find the following:

1. The Velocipede (4);
2. The Crankshaft (4);
3. The Steam Pump (4);
4. The Industrial Revolution (4).

Here, we find a wide range of time periods, from 1206 (the Crankshaft), up to the Velocipede (1817). This set of inventions includes both energy- and information-, but no material-handling systems.

It is also interesting to examine the other end of the spectrum and consider the lower scoring inventions. *Low* here, as I've pointed out already, is a relative term. The weakest invention in this series, in terms of creativity scores, is 11 out of 16 (Standardised Coinage). On average, therefore, this still represents a score of 2.75 out of 4 in each criterion—in absolute terms, these are not *uncreative* scores! Although, I suggested that a criterion for being in this book was that everything should be completely relevant and effective, it turns out that was not the case. Some of our inventions, as we have seen, received slightly *imperfect* scores for this criterion. There seemed to be good reasons for that. Even though the invention might have been in advance on preceding solutions, and may even have been important in stimulating later inventions, some clearly could have done what they were supposed to do somewhat better than was the case. The lowest scores for relevance and effectiveness that any of our inventions achieved is 2.5 out of 4. In other words, none of the inventions completely failed to solve the target problem! Most did a very good job, and a handful were a little imperfect (but not terrible). The least relevant and effective solutions in our analysis are, jointly

1. Antisense (Gene) Therapy (2.5);
2. Standardised Coinage (3);
3. The Wright Flyer (3);
4. The Scythe (3).

No particular pattern is evident here. The inventions span a wide range of time periods and include energy- and information-handling systems. Combining effectiveness with elegance to represent functional solutions, we once again find a perfect correlation between the two. The same four inventions that score lowest for effectiveness, score lowest when we add in elegance. In each case, however, this reinforces the fact that, had they been better executed, these inventions would have scored higher.

The least novel—i.e. the least *surprising*—solutions in our analysis are, jointly

1. The Scythe (2.5);
2. Standardised Coinage (2.5);
3. Sputnik (2.5);
4. The Nuclear Power Plant (2.5).

Once again, a very wide range of time periods, and types of system give us no particular clues as to why an invention may or may not be as novel as possible. One conclusion we might draw here is that each of these solutions may have a strong degree of "well, that's obvious" about it.

The least elegant of our innovations—i.e. the least *pleasing*—are, jointly

1. The Hand Axe (2.5);
2. The Smallpox Vaccine (2.5);
3. The Steam Pump (2.5);
4. Standardised Coinage (2.5);
5. Antisense (Gene) Therapy (2.5).

When we imagine ourselves seeing and using these for the first time, it's not hard to speculate that we would have viewed them as useful, but a bit clunky. Each could have been executed better.

Finally, the solutions with the lowest genesis—the weakest *disruptors* or *paradigm breakers*—are, jointly

1. The Hand Axe (2);
2. Sewerage Systems (2);
3. The WWW (2).

Once again, placing ourselves in the era in which each of these was introduced, my reaction is that none of these evokes the same sense of "this changes everything!" that, for example writing, smartphones or skyscrapers would have. You might be horrified to apply that idea to the WWW, but keep in mind that this is a *relative* judgement. Another piece of personal evidence to support my claim is the following. Twenty or so years ago (namely, in the late 1990s), many people predicted the demise of face-to-face education. The Web would replace everything, and students would log on from home, and never again need to attend a physical school or university.

That simply has not happened. Of course, the WWW has added immense value and flexibility to education, but the basic premise remains largely one of the students gathering together with the teacher, interacting, discussing and collaborating. It is true that this is frequently *supplemented* by technology and the WWW. However, my experience is that the underlying paradigm is largely the same as it was pre-WWW. That is why I give it a lower score for genesis—it *added to* the paradigm, but it did not break it.

Before we try to draw any overarching conclusions, there are two more sets of scores to consider. First, how do the inventions compare when we look across the categories of energy-handling-, material-handling- and information-handling systems. Although we have slightly different numbers in these categories, the following average scores for total creativity (out of 16) are

- Energy-handling systems (14 in total) = 13.43;
- Material-handling systems (8 in total) = 13.00;
- Information-handling systems (8 in total) = 12.50.

It is difficult to draw any firm conclusions from this. The differences in average scores are reasonable—about half a point in each case—but without a larger sample, and without more independent scores (i.e. someone other than me), it's risky to conclude that there is any significant difference in creativity on the basis of the type of system. Are information-handling systems inherently less creative than energy-handling systems, or is that just a reflection of the inventions chosen, and even my personal biases and preferences?

Finally, we can also look at creativity scores by our different time periods, and we find the following average scores (again, out of 16):

- Prehistory (3) = 12.83;
- The Classical Period (3) = 11.83;
- The Dark Ages (3) = 13.67;
- The Renaissance (3) = 12.33;
- The Age of Exploration (3) = 14;
- The Age of Enlightenment (3) = 13.67;
- The Romantic Period (3) = 13.67;
- The Modern Age (3) = 13.17;
- The Space Age (3) = 12.50;
- The Digital Age (3) = 13.33.

Although I am also reluctant to declare any firm conclusions from these figures, it appears, from our limited sample, that we have something of a clear winner in terms of the different time periods. Indeed, there is something of a bulge around the period from 1600 through 1900, peaking in the Age of Exploration. If we were to try and extract a conclusion from this, I would be tempted to say that this reflects the driving force of humankind's expansion beyond familiar borders, equipped with a new attitude to knowledge, and driven by a range of practical problems that arose from this new-found curiosity about the external world.

As we draw our exploration of humankind's creativity and problem-solving ability to a close, what, if anything, can we conclude from the inventions that we have analysed, and the range of problems that we have seen addressed? For me, the overwhelming conclusion is one that I've already hinted at, and one that I find very encouraging. It's very much a reaffirmation of the idea behind the title of the book. From the earliest days of our anatomically modern ancestors, right up to the present, we have a remarkable capacity to recognise problems and solve them, using whatever knowledge, skills and technology are at our disposal. These solutions are not always the most effective they can be, or the most novel, elegant or disruptive, but through this remarkable capacity that we possess—the desire and motivation to harness our mental capacity to generate new and effective solutions to the problems we face—we humans have thrived on a planet that has thrown, and continues to throw, everything it can at us. Some of these problems are of our own making, while many are just the natural consequences of a constantly changing environment. Regardless of the source, our status as *Homo problematis solvendis* means that we have all the skills and abilities we need to thrive, provided we nurture our creativity, and put it to good use.

Appendix A
The Creative Solution Diagnosis Scale (CSDS)

The Creative Solution Diagnosis Scale (CSDS)[1] is a measurement scale designed to help people assess the creativity of artefacts. Those artefacts can be any outcome of an activity: for example, written essays, poems, constructions, artefacts. Equally, the artefact may be any solution to a problem.

The CSDS is designed to assist in the process of understanding what makes something (a solution, or *product*) creative. A key concept from research is that the creativity of solutions is not just defined by something that is new, or different, but also something that is effective, and relevant to the problem or task at hand.

We also know that it can be hard for people to recognise and judge creativity. Experts are pretty good at doing this in their field of expertise, but what about other people, e.g. outside of a given field? The CSDS is designed to help anybody judge the creativity of anything, and to do so in a rigorous and systematic fashion. In this way, the CSDS can also be used, for example by teachers as a means of giving students formative feedback on their work, when creativity is a focus.

To do this, the CSDS uses a number of indicators grouped into five categories.

The five main categories that define the creativity of an artefact are:

- Relevance and Effectiveness—*the artefact is fit for purpose*;
- Problematisation—*the artefact helps to define the problem/task at hand*;
- Propulsion—*the artefact sheds new light on the problem or task*;
- Elegance—*the artefact is well executed*;
- Genesis—*the artefact changes how the problem/task is understood*.

However, these can still be hard for people to recognise in an artefact without some further guidance. Therefore, we have developed 21 more detailed *indicators* of creativity—these are easily recognisable characteristics of an artefact that helps define creativity. For example:

---

[1]See also, Cropley, D. H. and Cropley, A. J. (2016). Promoting creativity through assessment: A formative CAA tool for teachers, *Educational Technology Magazine, 56:6*, pp. 17–24.

© Springer Nature Singapore Pte Ltd. 2019
D. H. Cropley, *Homo Problematis Solvendis—Problem-solving Man*,
https://doi.org/10.1007/978-981-13-3101-5

- Performance = "the artefact does what it is supposed to do".

To make an assessment of the creativity of an artefact a person goes through the 21 indicators and rates the artefact on a scale from "0" (that indicator does not apply to the artefact), through to "4" (that indicator applies very much to the artefact). By combining these ratings together, we get a very detailed assessment of the artefact's creativity, and one that can be used for formative assessment—to help students understand what to do to increase the creativity of the things that they produce.

The full CSDS assessment scale is shown below:

A. **Relevance and Effectiveness** (the artefact is fit for purpose):

  a. CORRECTNESS (the artefact accurately reflects conventional knowledge and/or techniques);
  b. PERFORMANCE (the artefact does what it is supposed to do);
  c. APPROPRIATENESS (the artefact fits within task constraints).

B. **Problematisation** (the artefact helps to define the problem/task at hand):

  a. DIAGNOSIS (the artefact draws attention to shortcomings in other existing artefacts);
  b. PRESCRIPTION (the artefact shows how existing artefacts could be improved);
  c. PROGNOSIS (the artefact helps the observer to anticipate likely effects of changes).

C. **Propulsion** (the artefact sheds new light on the problem/task):

  a. REDIRECTION (the artefact shows how to extend the known in a new direction);
  b. REINITIATION (the artefact indicates a radically new approach);
  c. REDEFINITION (the artefact helps the observer see new and different ways of using the artefact);
  d. GENERATION (the artefact offers a fundamentally new perspective on possible artefacts).

D. **Elegance** (the artefact is well executed):

  a. CONVINCINGNESS (the observer sees the artefact as skilfully executed, well finished);
  b. PLEASINGNESS (the observer finds the artefact neat, well done);
  c. COMPLETENESS (the artefact is well worked out and "rounded");
  d. GRACEFULNESS (the artefact is well proportioned, nicely formed);
  e. HARMONIOUSNESS (the elements of the artefact fit together in a consistent way).

E. **Genesis** (the artefact changes how the problem/task is understood):

  a. FOUNDATIONALITY (the artefact suggests a novel basis for further work);

b. TRANSFERABILITY (the artefact offers ideas for solving apparently unrelated problems);

c. GERMINALITY (the artefact suggests new ways of looking at existing problems);

d. SEMINALITY (the artefact draws attention to previously unnoticed problems);

e. VISION (the artefact suggests new norms for judging other artefacts existing or new);

f. PATHFINDING (the artefact opens up a new conceptualisation of the issues).

In this book, the readers will note that I have used only four criteria to assess each invention. The only difference in the scales is that I frequently combine *problematisation* and *propulsion* together as a single criterion of *novelty*. Aside from that, in making my assessments of creativity for each invention, I used the 21 indicators listed, and calculated a single score, to the nearest half point, for each of the four main criteria.

Appendix B
Table of Creativity Scores

The following table lists the 30 inventions in chronological order, with the details of their creativity scores broken down by category (Table 1).

Table 1 Creativity scores for the 30 inventions

Invention	Effectiveness	Novelty	Elegance	Genesis	Total
The Hand Axe	4	3	2.5	2	11.5
Oars	4	3.5	3	2.5	13
Cuneiform	3.5	3.5	3.5	3.5	14
Standardised Coinage	3	2.5	2.5	3	11
The Construction Crane	3.5	3	3	3	12.5
The Julian Calendar	3.5	3	3	2.5	12
Paper	4	3.5	3	2.5	13
The Spinning Wheel	4	3	3	3	13
The Crankshaft	4	3.5	3.5	4	15
The Scythe	3	2.5	3	2.5	11
The Printing Press	4	3	3	2.5	12.5
The Condom	4	3.5	3	3	13.5
The Slide Rule	4	3	4	3	14
The Pendulum Clock	3.5	4	4	3	14.5
The Steam Pump	3.5	3.5	2.5	4	13.5

(continued)

© Springer Nature Singapore Pte Ltd. 2019
D. H. Cropley, *Homo Problematis Solvendis—Problem-solving Man*,
https://doi.org/10.1007/978-981-13-3101-5

Table 1 (continued)

Invention	Effectiveness	Novelty	Elegance	Genesis	Total
The Leyden Jar	3.5	3.5	3	3.5	13.5
The Industrial Revolution	4	4	3.5	4	15.5
The Smallpox Vaccination	3.5	3	2.5	3	12
The Velocipede	4	4	4	4	16
Sewerage Systems	4	3	3.5	2	12.5
The Electric Lightbulb	4	3	3	2.5	12.5
Steel-Girder Skyscrapers	4	3.5	4	3	14.5
The Wright Flyer	3	3	3	2.5	11.5
Antibiotics	4	3.5	3	3	13.5
Sputnik I	4	2.5	3.5	2.5	12.5
Nuclear Power	4	2.5	3.5	2.5	13
Antisense (Gene) Therapy	2.5	3.5	2.5	3.5	12
The WWW	4	3	4	2	13
Artificial Skin	3.5	3	3	3.5	13
Smartphones	4	3	3	3.5	13.5

Bibliography

The following texts are enormously readable, and have provided excellent support for my research.

1. Bernstein, P. L. (1996). *Against the Gods: The remarkable story of risk.* New York, NY: John Wiley and Sons, Inc.
2. Brandt, A, and Eagleman, D. (2017). *The Runaway Species: How human creativity remakes the world.* Edinburgh, UK: Canongate Books Ltd.
3. Cadbury, D. (2012). *Seven Wonders of the Industrial World.* London, UK: Harper Perennial.
4. Diamond, J. (1999). *Guns, Germs, and Steel: The fates of human societies.* New York, NY: W. W. Norton & Company, Ltd.
5. Dickson, P. (2001). *Sputnik: The shock of the century.* London, UK: Walkcr Publishing Company.
6. Gertner, J. (2012). *The Idea Factory: Bell Labs and the great age of American innovation.* London, UK: The Penguin Press.
7. Harari, Y. N. (2011). *Sapiens: A brief history of humankind.* London, UK: Vintage Books.
8. Holmes, R. (2010). *The Age of Wonder.* New York, NY: Vintage Books.
9. Khan, A. (2017). *Adapt: How we can learn from nature's strangest inventions.* London, UK; Atlantic Books.
10. Kirby, R. S., Withington, S., Darling, A. B., & Kilgour, F. G. (1990). *Engineering in History.* New York, NY: Dover Publications, Inc.
11. Petroski, H. (1996). *Invention by Design: How engineers get from thought to thing.* Cambridge, MA: Harvard University Press.
12. Schilling, M. A. (2018). *Quirky.* New York, NY: PublicAffairs.

© Springer Nature Singapore Pte Ltd. 2019
D. H. Cropley, *Homo Problematis Solvendis—Problem-solving Man,*
https://doi.org/10.1007/978-981-13-3101-5

Printed in the United States
By Bookmasters